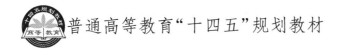普通高等教育"十四五"规划教材

# 有机化学实验

(Experimental Organic Chemistry)

主　编　潘世光
副主编　臧洪俊　王凤勤　张建新
　　　　于鹏飞　吕　义

中国石化出版社

## 内 容 提 要

有机化学实验教学的主要目的是培养学生实验的操作技能和动手能力。本书围绕"以问题为核心,让学生在实验中成长"进行编写,实验课程体系设计从低到高,从基础到前沿,从基本技能训练到综合能力培养,层次分明,逐级提高。全书采用英文编写,结合了国内外有机化学实验教学的实际需要。全书分为四部分:有机化学实验的一般知识,实验操作的基本技能,有机化合物的制备(22 个基础实验和 1 个综合设计性实验)和附录,涵盖了化学、化工等专业教学大纲中对有机化学实验课的基本要求。每个实验包括实验目的、实验原理、仪器试剂、实验步骤、注意事项和思考题等部分。书后附录收集了各类实验所需的数据表,便于学生及相关人员查阅使用。

本书可作为普通高校化学、化工等专业以及相近专业本科生的双语教材,也可供相关专业的科研人员使用。

### 图书在版编目(CIP)数据

有机化学实验 = Experimental Organic Chemistry:英文 / 潘世光主编. —北京:中国石化出版社,2022.8

普通高等教育"十四五"规划教材
ISBN 978-7-5114-6772-0

Ⅰ. ①有… Ⅱ. ①潘… Ⅲ. ①有机化学-化学实验-高等学校-教材-英文 Ⅳ. ①O62-33

中国版本图书馆 CIP 数据核字(2022)第 131611 号

未经本社书面授权,本书任何部分不得被复制、抄袭,或者以任何形式或任何方式传播。版权所有,侵权必究。

中国石化出版社出版发行
地址:北京市东城区安定门外大街 58 号
邮编:100011　电话:(010)57512500
发行部电话:(010)57512575
http://www.sinopec-press.com
E-mail:press@sinopec.com
北京科信印刷有限公司印刷
全国各地新华书店经销

\*

710×1000 毫米 16 开本 8.75 印张 154 千字
2022 年 9 月第 1 版　2022 年 9 月第 1 次印刷
定价:36.00 元

# 前　言

随着国家"一带一路"倡议的提出，我国与多个国家和地区人文交流机制的建立，中华优秀传统文化、中国高等教育质量、稳定安全的社会环境、广阔多元的发展空间、中文在国际交流中的通用程度等，都在不断吸引世界青年选择留学中国。外国学生来国内留学是我国教育事业的重要组成部分，是我国教育对外开放和对外形象展示的窗口，是中外友好交往、促进民心相通的纽带，是讲好中国故事的平台，是"一带一路"人才培养的重要方式，也是向世界会聚人才、助力科研国际交流的重要渠道。

随着来华留学生的与日俱增，高校对外国留学生实行双语教学已经成为常态；同时，随着有机化学实验在教学内容、方法和手段上不断更新，原有的实验教材已不能满足和适应来华留学生的教学和学习需求。因此，为了配合留学生双语教学的需求，同时为了能够培养留学生具有系统扎实的基础知识、实验技能，又具有一定的动手能力和思考能力，我们根据教育部关于化学、化工等专业以及相近专业"有机化学实验"教学内容，并结合教学团队多年的教学经验，参考国内、外有关实验教材和参考书，组织编写了本书。本书主要是作为有机化学课程的实验室指南，指导学生在实验室中学习。

本书前两章立足于基础实验内容，包括一般基础知识、常用仪器的使用方法以及实验操作技能等，培养留学生通过科学实验研究问题的基本意识和思维习惯，确保留学生能够规范、正确、熟练地进行基本实验操作；同时，还包括实验室的安全和注意事项等。第3章包含22个基础实验和1个综合设计性实验，详细介绍了各个实验的实验目的，基本原理与实验方法、实验内容与步骤等内容，力求繁简适当、通俗易懂。在综合设计性实验中，引入有机-无机杂化催化材料的新概念和新技术，促进留学生对绿色有机合成方法的理解并让留学生系统地接触科研前沿，提升研究兴趣。第4章附录收集了各类实验所需的数据表，便于留学生查阅使用。

本书采用英文编写，内容从简到繁、由浅至深，选择力求做到理论与实践结合、基础与前沿结合、个体与群体结合，体现基础性，突出综合性，加强实用性和趣味性。

由于作者水平有限，编写时间仓促，书中难免会出现一些错误或者不准确的地方，恳请同行及读者批评指正。任何关于改进实验的建议，我们也将欣然接受。

编者

2022 年 9 月

# Contents

**Chapter 1 General Knowledge of Organic Chemistry Experiments** ( 1 )
  1.1 Laboratory rules ( 1 )
  1.2 Prevention and treatment of experimental accidents ( 2 )
  1.3 Experiment record and experiment report ( 7 )
  1.4 Check items if the experiment does not give results ( 9 )
  1.5 Glassware commonly used in organic chemistry laboratories ( 10 )
  1.6 Methods of cleaning, drying, and maintaining the instruments ( 15 )
  1.7 Application of low-temperature refrigeration ( 19 )
  1.8 Commonly heating appliances and equipment ( 19 )

**Chapter 2 Basic Operation and Experimental Technology of Organic Chemistry Experiment** ( 23 )
  2.1 Determination of physical constants of organic compounds ( 23 )
  2.2 Simple glazier operation ( 28 )
  2.3 Cooling and heating ( 31 )
  2.4 Separation and purification of organic compounds ( 34 )
  2.5 Drying and the use of desiccant ( 60 )

**Chapter 3 Experiments** ( 64 )
  3.1 Distillation of saturated aqueous solution of $n$-butanol ( 64 )
  3.2 Synthesis of 1-bromobutane ( 68 )
  3.3 Synthesis of ethyl acetate ( 72 )
  3.4 Synthesis of $n$-butyl acetate ( 75 )
  3.5 Synthesis of ethyl benzoate ( 77 )
  3.6 Synthesis of $n$-butyl ether ( 81 )
  3.7 Synthesis of acetanilide ( 83 )
  3.8 Recrystallization of acetanilide ( 85 )
  3.9 Synthesis of $p$-bromoacetanilide ( 88 )
  3.10 Synthesis of benzyl alcohol and benzoic acid ( 90 )

| | | |
|---|---|---|
| 3.11 | Synthesis of α-furyl methanol and α-furoic acid | ( 93 ) |
| 3.12 | Synthesis of benzyl alcohol | ( 95 ) |
| 3.13 | Synthesis of benzoic acid | ( 97 ) |
| 3.14 | Synthesis of cyclohexene | ( 99 ) |
| 3.15 | Synthesis of cyclohexanone | (101) |
| 3.16 | Synthesis of cinnamic acid | (104) |
| 3.17 | Synthesis of acetylsalicylic acid | (106) |
| 3.18 | Synthesis of 2-methyl-2-butanol | (108) |
| 3.19 | Synthesis of *tert*-butylhydroquinone | (112) |
| 3.20 | Synthesis of nitrobenzene | (114) |
| 3.21 | Synthesis of adipic acid | (117) |
| 3.22 | Synthesis of dimethyl adipate | (119) |
| 3.23 | Comprehensive experiment: green synthesis of 2,2′-furil | (121) |

**Chapter 4  Appendix** ...... (127)

| | | |
|---|---|---|
| 4.1 | List of the element with their symbols and atomic masses | (127) |
| 4.2 | Structures and names for common solvents used in organic chemistry | (131) |
| 4.3 | Properties for common solvents used in organic chemistry | (132) |
| 4.4 | Cooling baths | (132) |
| 4.5 | Common azeotropes | (133) |
| 4.6 | Pressure-temperature nomograph | (134) |
| 4.7 | Periodic table | (134) |

# Chapter 1

# General Knowledge of Organic Chemistry Experiments

Organic chemistry experiment is an important basic experiment course. The purpose of this course is to train students to grasp the experiment's basic operation skills and basic knowledge firmly. Experiments can strengthen students' understanding of the fundamental theories and concepts of organic chemistry, train students to correctly choose organic compounds, and master the general synthesis, separation, purification, and identification methods. After studying, students can analyze problems, solve problems, and innovative thinking methods, with the preliminary ability to engage in scientific research; to train students to combine theory with practice, seek truth from facts, scientific attitude, and good working habits.

## 1.1 Laboratory rules

In order to ensure the smooth progress of the experimental class and enable students to develop a good laboratory work style, require students to abide by the following organic chemistry laboratory rules:

( ⅰ ) Pre-study carefully before the experiment, understand the experimental purpose, principle, synthetic route and possible problems in the experimental process, and safety matters that should be paid attention to, write a pre-study outline and check the physical and chemical properties of the relevant compounds.

( ⅱ ) Familiar with the switch positions of water, electricity, gas in the laboratory, and the use of fire-fighting equipment, and master the basic knowledge for preventing fire, poisoning, explosion, and first aid.

( ⅲ ) During the experiment, strictly follow the operating procedures, wear safety glasses and lab coats for experimentation. Observe experimental phenomena and record faithfully. The experimental reagents shall not be discarded or lost randomly. Always keep the personal experiment table, the ground, and the public experiment table clean and tidy during the experiment. Follow the teacher's guidance, obey laboratory disciplines, keep quiet, and do not leave the laboratory without authorization during the experiment.

(ⅳ) The use of flammable and explosive reagents should be kept away from fire sources, and take safety precautions according to the accidents. Smoking or eating food is strictly prohibited in the laboratory, and all reagents are not allowed to taste. After the experiment, wash your hands carefully.

(ⅴ) Take care of laboratory equipment, conserve the amount of reagent, water, electricity, and gas; strictly prevent mercury and other toxic compounds pollution, promptly report accidents such as thermometer damage, and take emergency measures under the guidance of teachers. It is strictly forbidden to pour waste acids, waste alkalis, and solids into the sink. When damaging the laboratory equipment, you should truthfully describe the damage.

(ⅵ) When you finish the experiment, you need to submit your experiment record to teachers for review and signature. The experimental data and results should be recorded realistically and do not arbitrarily modify, forge or copy others' experimental data.

(ⅶ) On-duty students are responsible for cleaning the laboratory, checking and closing the water, electricity, and gas valves. You can leave the laboratory only after the teacher's inspection and approval.

## 1.2 Prevention and treatment of experimental accidents

Many kinds of reagents are used in organic chemistry experiments, and most of them are flammable, explosive, highly toxic, and corrosive. Improper use may cause fire, poisoning, burns, explosions, and other accidents. Most of the instruments used in the experiment are glass products coupled with gas, electrical equipment, which increase the potential risk. But, if we adopt appropriate precautions, experimenters have the basic knowledge of the experiment, pay attention to safe operation, master the correct operating procedures, and abide by the organic experiment rules, accidents can be avoided entirely.

### 1.2.1 Understand the familiarity of the laboratory set-up

When entering the laboratory, the experimenter must first understand and be familiar with the positions of the laboratory's electric switch, gas switch, water switch, and safety equipment such as fire extinguishers, sandboxes, asbestos cloth, and know how to use them. Do not move the location of safety equipment at any time.

### 1.2.2 Check

Before starting the experiment, carefully check whether the instrument is damaged and whether the device is correct and stable.

### 1.2.3 Fire prevention

Flammable solvents commonly used in laboratories, such as ethanol, ether, petroleum ether, benzene, toluene, acetone, ethyl acetate, and other flammable liquids, should not be heated in an open container, and the correct heating method should be used according to the nature of the solvent.

Flammable organic solvents, especially low boiling point flammable solvents, at room temperature that has a considerable vapor pressure, when the air mixed with flammable organic solvent vapor reaches a certain limit, in the presence of open fire (or even sparks generated by electrical switches, or due to static friction, knocking sparks) will explode. The density of organic solvent vapor is greater than that of air and will drift along with the table or floor or be deposited in low places, so do not pour flammable solvents into waste tanks. Laboratory refrigerators should not be stored in excess of flammable organic solvents to prevent a large area from catching fire and exploding due to sparking in the refrigerator. When distilling flammable solvents, the device should prevent flammable vapors from leaking, and the receiver branch should be connected to a rubber tube so that the residual gas is discharged down the sink. When required, operate in a fume hood. Remember: Heating volatile liquids should be kept away from open flames and as far as possible without open flames, which is the most basic principle of fire prevention. If open flames must be used, attention should be paid to choosing a suitable heating bath. If an open flame must be used, care should be taken to select a suitable heating bath, an oil bath, a water bath, an asbestos net depending on the boiling point of the reaction solution.

### 1.2.4 Never heat a closed system

Operating at atmospheric pressure, the instrumentation needs to have a vent to the atmosphere in the apparatus, never heat a closed system as this will cause the increase of its system pressure and lead to an explosion.

### 1.2.5　Fire management

In the event of a fire, do not panic, first turn off the gas immediately, pull down the electric switch, cut off the power supply and quickly remove any flammable materials around the scene of the fire. Usually, water is not used to extinguish the fire to prevent the chemical from reacting with water and causing a bigger accident. If the solvent in the apparatus is on fire, it is best to use a large piece of asbestos cloth to cover the fire and prevent the use of sand and soil to put out the fire, which may break the glass apparatus and cause a wider spread of the fire. Small fires can be extinguished with a wet cloth or asbestos cloth, but if the fire is large, the following extinguishing equipment should be used:

(a) Carbon dioxide fire extinguisher

A fire extinguisher commonly used in organic laboratories, the steel cylinder is filled with compressed liquid carbon dioxide, which is ejected when the switch is turned on to extinguish fires in organic and electrical equipment. The correct method of operation is to hold the fire extinguisher in one hand, and the other hand should be held in the handle of the $CO_2$ spraying horn barrel, do not hold your hand on the horn barrel, because with the spraying of carbon dioxide, the pressure drops suddenly, the temperature also drops suddenly, hand on the horn barrel will be frostbite.

(b) Carbon tetrachloride (CTC) fire extinguisher

Used to extinguish fires in or near electrical appliances. Not suitable for use in small and poorly ventilated laboratories as CTC produces highly toxic phosgene when extinguished at high temperatures. It is not suitable for use in the presence of sodium metal as the reaction between carbon tetrachloride and sodium metal will cause an explosion. To use this extinguisher, the nozzle is continuously pumped and the CTC is ejected from the nozzle.

Regardless of the type of extinguisher used, the fire should be extinguished starting from the perimeter of the fire and working towards the center. If your clothes are on fire, do not panic and run around, causing the flames to expand, you should quickly take off your clothes to put out the fire, or use a thick coat, or asbestos cloth to extinguish the fire. In serious cases, you should immediately lie on the ground (so that the flames do not burn to your head) and roll to smother the fire, or turn on the tap to extinguish with water.

### 1.2.6 Skin burn management

Skin burns can be caused by contact with corrosive chemicals (strong acids, bases, bromine, etc.) and should be treated separately according to the following conditions, and serious cases should be taken to hospital immediately.

(a) Concentrated acid burn

Rinse immediately with plenty of water, wash with 3% to 5% sodium bicarbonate solution, and paint scald ointment.

(b) Bromine burn

Burns caused by bromine are particularly serious and should be washed immediately with plenty of water and then scrubbed with alcohol until free of bromine, followed by the application of cod liver oil ointment.

(c) Concentrated alkali burn

Rinse immediately with plenty of water, wash with a 1% to 2% boric acid solution, and finally wash again with water and paint oil paste.

### 1.2.7 Protects against liquid splashes

Wear protective eyewear during experiments and do not splash corrosive chemicals or hot solvents and reagents into your eyes. When measuring chemicals, place the measuring cylinder on the laboratory bench and add the liquid slowly, without getting close to the eyes. Once splashed, immediately rinse with plenty of water and take to hospital promptly.

### 1.2.8 Prevention of cuts

Cuts are a frequent accident in the laboratory. They often occur when pulling glass tubes or setting up instruments. When a cut is made, the first thing to do is to check the wound for glass shavings. If any, remove them, wash the wound with water, paint iodine, and wrap it with gauze. Do not expose the wound to chemicals that can cause poisoning.

### 1.2.9 Protection against toxic reagents

Wear rubber gloves and protective glasses when working with toxic chemicals (e.g. benzene, nitrobenzene, benzidine, nitroso compounds, etc.) and corrosive chemicals. For volatile and toxic reagents, always operate in a fume hood when using them. No chemicals should be tasted.

### 1.2.10  Beware of electric shock

When using electrical appliances, the body should be prevented from direct contact with the conductive parts of the apparatus, and the electrical plug should not be touched with wet hands or wet objects held in the hands. To prevent electric shock, the metal casing of the apparatus and equipment should be connected to an earth wire, and the experiment should be completed by switching off the apparatus power switch and then unplugging it. In case of electric shock, the power supply should be cut off immediately or use a non-conductive object to disconnect the electrocution victim from the power supply, and then apply artificial respiration to the electrocution victim and send him/her to hospital immediately for rescue.

### 1.2.11  Poisoning

Rinse the mouth immediately with plenty of water if the chemical is splashed into the mouth. If the chemical is swallowed by mistake, an antidote should be given according to the nature of the poison, and take to hospital immediately.

(a) Corrosive poisons

For strong acids, drink plenty of water and then take aluminium hydroxide paste and egg-white; for strong bases, also drink plenty of water, then take vinegar, sour juice, egg-white. For both acid and alkaline poisoning, infuse milk and do not take vomiting agents.

(b) Poisons with nerve stimulating properties

Drink a large quantity of milk or egg-white immediately to dilute and relieve, then a large spoonful of magnesium sulfate (about 30g) dissolved in a glass of water to induce vomiting. Sometimes a finger can be put down the throat to induce vomiting and take to hospital immediately.

For poisoning by inhalation of toxic gases, immediately remove the poisoned person to the outside, unbutton and untie the collar and buttons of the coat, and give treatment according to the type of gas inhaled. For example, for small amounts of chlorine or bromine, rinse the mouth with a sodium bicarbonate solution.

To deal with incidents, the laboratory should have a first aid kit with some of the following items:

( ⅰ ) Bandages, gauze, desiccated cotton, ointment, medical forceps, scissors, etc.

( ii ) Acetic acid solution (2%), boric acid solution (1%), sodium bicarbonate solution (1% and saturated solution), medical alcohol, glycerin, etc.

( iii ) Vaseline, band-aids, or tannin ointment, burn ointment and antiseptic etc.

## 1.3  Experiment record and experiment report

### 1.3.1  Preview and experiment records

The purpose of the experiment, the requirements, and the physical and chemical properties of the reagents and the products should be thoroughly studied before the experiment so that the whole experiment can be well understood, and a preview report should be written in the experiment record book. If you do not prepare for the experiment, you will not achieve the expected results.

The preview report is part of the lab record and is the basis for studying the content of the experiment and writing the experiment report. You can refer to the following items to do the preview report before the experiment starts.

( i ) Name of the experiment, purpose and requirements of the experiment, reaction formula (target reaction and side reactions).

( ii ) Physical and chemical constants of reagents and products (relative molecular mass, properties, refractive index, density, melting point, boiling point, and solubility).

( iii ) Following the reaction equation, calculate the theoretical amount of reactants used and the theoretical yield of the product.

( iv ) The type and model size of the instrument used.

( v ) Brief operation steps.

The content that has been involved in the preview report will be further understood and updated during the experiment. Each page of the experiment record book can be divided into two columns. On the left, write the preview content, and on the right of the corresponding column, write the updated knowledge and supplements in the experiment and the observed experimental phenomenon. Each column must be recorded in permanent ink, and each page of the record book must be marked with the date and page number.

After the experiment starts, it is essential to make a record of actual observations. The start time of the experiment, the whole process of the experiment, such as the change of reaction temperature, whether the reaction is exothermic, color change, whether there is crystallization or precipitation, especially the opposite phenomenon should be paid special attention to, and the observed phenomena it should be accurately recorded in the experiment record book, because it will be of great help to interpret the experimental results correctly.

Preparing the experiment and keeping a record of the experiment is extremely important for guiding the operation of the experiment, deeply understanding the content of the experiment, and effectively using the experiment time.

(a) Experimental records during the reaction

All phenomena observed (detailed description).

The sequence of each step and the time it takes.

Track the experiments and results of the reaction process.

(b) After the reaction is over, record the follow-up work in time

Method of post-processing operation.

Product purification method.

Yield and percentage yield.

Product purity inspection methods and results (including appearance and test data), such as "slightly yellow needles, melting point 94.5-96.2℃" "GC, retention value 1.26min, normalized treatment, 95.2%", etc.

Structural analysis results of the product.

## 1.3.2 Experiment report

The experimental report is a written reflection of the experimental results obtained by the experimenter after completing the experiment. Writing an experimental report is a practical activity for the experimenter to learn the scientific research methods of organic chemistry. Writing an experimental report is a process for the experimenter to review the experiment, summarize the operational experience and re-learn. It can also be a process of deepening awareness and gaining a deeper understanding of theory. From the experimental report, teachers can comment on the experimenter's understanding of experimental theory and experimental phenomena, the ability to express themselves in writing, and the responsibility of the experiment. Therefore, writing a standard, accurate and complete laboratory report is one of the basic requirements of organic chemistry laboratory courses.

The experimental report should be guided by the experimental principles, based on the experimental records, and uniformly written in a standardized format using the units and symbols specified in GB 3101—93 "General Principles of Relevant Quantities, Units, and Symbols". Except for the experimental set-up diagram, all other content should be written in pen, not pencil. The writing should be neat and not scribbled. Diagrams should be standardized and not sketched. All items in the experimental report should be written without gaps. This part of the work is to be completed after class.

The written content of the organic chemistry experiment report can be roughly divided into the following items:

( i ) The experiment purpose.
( ii ) The reaction principle.
( iii ) The physical constants of the main reagents and products.
( iv ) The dosage and specifications of the main reagents.
( v ) Instruments.
( vi ) Experimental procedures and phenomena.
( vii ) Yield calculation.
( viii ) Experimental discussion.

## 1.4 Check items if the experiment does not give results

If your reaction does not proceed, ask yourself these questions:
When setting up the device:
( i ) Are you sure that the glassware used in the reaction is clean and the solvent is pure?
( ii ) Is your calculation correct?
( iii ) Are your starting materials pure?
( iv ) Is your reaction performed at the correct concentration?
When operating reaction:
( i ) Can you accurately monitor the reaction?
( ii ) Are you monitoring the reaction from the beginning to the end? When did the problem occur? (For example: do you let the reaction last so long that it broke down?)
( iii ) Do you know the temperature of the reaction mixture during the reaction?
( iv ) Have you added the reaction ingredients in the correct order and at the correct speed?
( v ) Is the stirring effective? If you notice a problem with mixing, please see the notes on mixing for some tips.

# 1.5 Glassware commonly used in organic chemistry laboratories

## 1.5.1 Introduction to glassware

(a) Ordinary glassware

The common glasswares are shown in Fig 1-1.

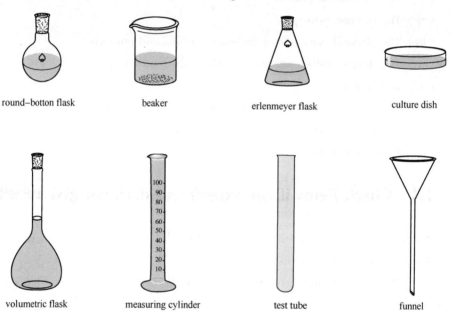

Fig 1-1 Schematic diagram of ordinary glassware

(b) Standard grinding-mouth glassware

In addition to the common glassware usually used in organic chemistry experiments, many glasswares with standard grinding mouths are also used, which have the characteristics of standard, general, series, etc. When the instruments are combined, they do not need a cork or rubber stopper, just with the help of the same number of internal and external grinding mouths. Different kinds of glasswares with the same grinding mouth can be combined at will, making the instruments easy to assemble and flexible to disassemble and prevent the reactants and products from being contaminated by the stopper.

The grinding mouths of glasswares are all cone-shaped, with a big end and a small end. The standard grinding mouth adopts the internationally accepted 1/10 taper, that is, for every 10 unit length of the grinding mouth, the diameter of the small end is reduced by

1 unit than the diameter of the large end. Commonly used standard grinding mouth numbers include 10, 14, 19, 24, 29, 34, 40, 50, etc. The number refers to the millimeter (mm) of the large end diameter of the grinding mouth cone. For example, 10/30 means that the diameter of the large end of the grinding mouth is 10mm, the length of the grinding mouth is 30mm, and the grinding mouth number is written as $\phi 10$.

Standard grinding-mouth glassware used commonly include the following categories:

(i) Grinding-mouth container:

The schematic diagram of grinding-mouth containers are shown in Fig 1-2.

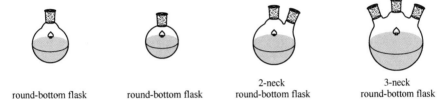

Fig 1-2  Schematic diagram of grinding-mouth containers

(ii) Grinding plugs and grinding knots:

The schematic diagram of grinding plugs and grinding are shown in Fig 1-3.

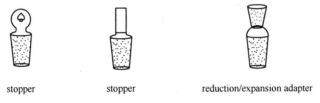

Fig 1-3  Schematic diagram of grinding plugs and grinding

(c) Condenser

Conventional synthesis experiments often use 200mm long condenser tube with a grinding-mouth of $\phi 19$ and schematic diagram are shown in Fig 1-4.

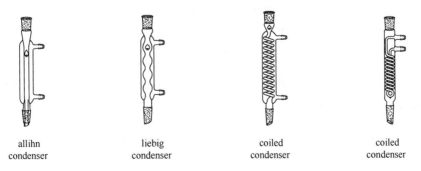

Fig 1-4  Schematic diagram of condensers

(d) Distillation head and receiving tube

The diagram of distillation head and receiving tube are shown in Fig 1-5.

still head      claisen adapter      vacuum distillation adapter

Fig 1-5  Schematic diagram of distillation head and receiving tube

(e) Other standard ground glass instruments

The schematic diagram of other standard ground glass instruments are shown in Fig 1-6.

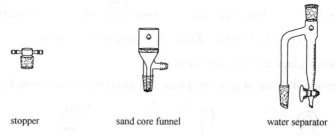

stopper      sand core funnel      water separator

Fig 1-6  Schematic diagram of other standard ground glass instruments

Glasswares should be handled gently, and pay attention to protect instruments with glass stoppers to prevent the stoppers from falling and breaking. Except for a few glassware such as test tubes, glassware cannot be directly heated using a fire. Do not use jars to store volatile organic solvents. For glasswares with pistons to be placed for a long time, a small piece of paper should be inserted between the grinding-mouth and the piston to prevent the plugs from sticking. If sticking occurs, add a small amount of organic solvent (glycerin or engine oil) to the gap of the grinding-mouth, and then heat it with a hair dryer to slowly infiltrate, or boiling it with water, and tap the stopper gently with a wooden block to pen it.

## 1.5.2  Pay attention to the following matters when using the grinding instrument

( i ) The grinding mouth must be clean and free of solid debris; otherwise the grinding mouth will not dovetail tightly resulting in air leakage. Hard solid particles also tend to damage the grinding mouth.

(ii) It should be disassembled and washed immediately after use. If left for a long time, it may cause the grinding joints to stick together and not be easily disassembled.

(iii) It is unnecessary to apply lubricants such as petroleum jelly to the grinding mouth to avoid contamination of the reactants and products. However, when strong alkalis are used in the reaction or high temperatures are used for heating, a little lubricant should be used to avoid sticking of the grinding mouth due to alkaline corrosion or high temperature; otherwise, it can not be disassembled easily. When distilling under reduced pressure, the grinding mouth should be coated with vacuum grease to prevent air leakage.

(iv) When installing the instrument, the connection angle between grinding mouths should be correct; otherwise, the grinding mouth will break due to the effect of the tilting stress.

### 1.5.3 Organic chemistry experiment instrument illustration

The schematic diagram of organic chemistry experiment instruments are shown in Fig 1-7 and Fig 1-8.

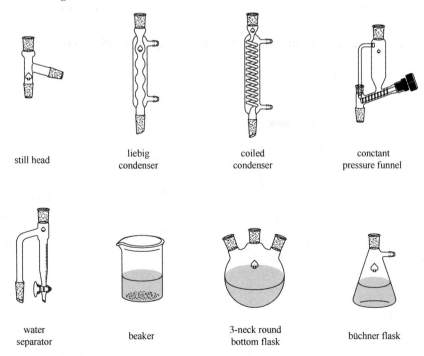

Fig 1-7  Schematic diagram of organic chemistry experiment instruments

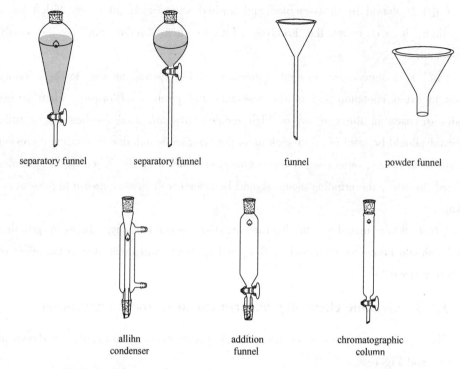

Fig 1-8  Schematic diagram of organic chemistry experiment instruments

## 1.5.4  Diagram of organic chemistry experiment device

The schematic diagram of diagram of organic chemistry experiment devices are shown in Fig 1-9, Fig 1-10 and Fig 1-11.

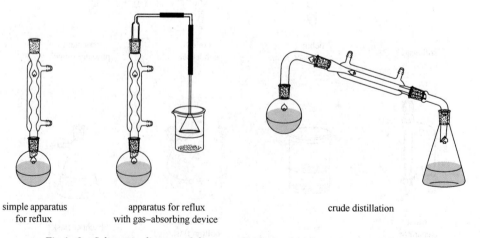

Fig 1-9  Schematic diagram of diagram of organic chemistry experiment devices

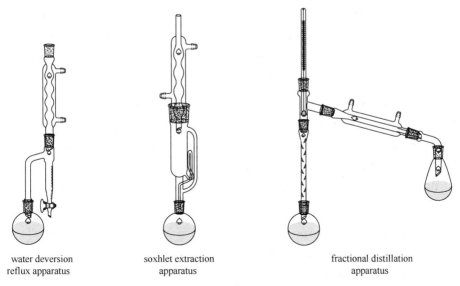

Fig 1-10 Schematic diagram of diagram of organic chemistry experiment devices, water deversion refiux, soxhle extraction and fractional distillation

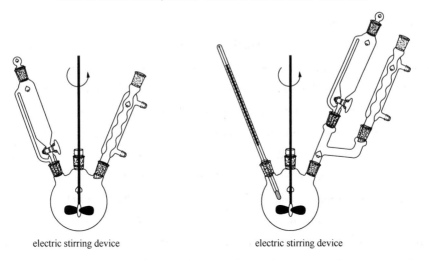

Fig 1-11 Schematic diagram of diagram of organic chemistry experiment electric stirring devices

# 1.6 Methods of cleaning, drying, and maintaining the instruments

## 1.6.1 Cleaning of common instruments

The experimental equipments used for organic reactions must be clean and dry. The sign of that the instrument is clean is: water flows down the wall, the inner wall is evenly wetted by the water, leaving a thin and uniform water film without water droplets.

To make cleaning easy and effective, it is best to clean the used instruments immediately after each experiment. Because the nature of the dirt was clear at the time of the experiment, it is easy to remove it with appropriate methods. If left for a long time, it will increase the difficulty of washing. In organic chemistry experiments, the simplest and most commonly used method for cleaning glass instruments is to use a long-handled brush (test tube brush) and decontamination powder to brush the walls of the apparatus until the dirt on the glass surface is removed, and then finally wash with water. However, when washing with a caustic solution, no brush is needed. You should not use sand when washing glassware as it can bruise or even break the glass. If it is difficult to clean, then wash with an appropriate detergent depending on the nature of the dirt. If the dirt is acidic (or alkaline), you can use an alkaline (or acidic) wash; organic dirt can be washed with an alkaline or organic solvent. Here are several commonly used lotions:

(a) Chromic acid

The lotion of chromic acid is highly oxidizing and very destructive to organic dirt. Pour out the water in the vessel, slowly pour the lotion of chromic acid, turn the vessel so that the lotion thoroughly soaks the unclean wall, and a few minutes later, pour the lotion back into the lotion bottle, and then rinse the vessel with water. If there is a small amount of carbonization residue on the wall, add a small amount of lotion, soak for a few minutes and then heat it using a small fire until bubbles escape, and the carbonization residue can be removed. However, when the lotion's color turns green, it is no longer effective and should not be poured back into the wash bottle.

(b) Lye and synthetic detergent

It can be prepared as a concentrated solution for washing grease and organic substances (such as organic acids).

(c) Hydrochloric acid

Concentrated hydrochloric acid can be used to wash away dirt such as manganese dioxide or carbonates that have adhered to the vessel's walls.

(d) Gelatinous or tar-like organic dirt

If it cannot be washed away by the above method, acetone, ether, or toluene can be used for soaking, but with a cover to avoid evaporation of the solvent. Alternatively, an ethanol solution of NaOH can be used as a detergent. Note that the solvent needs to be recycled and reused. You can use an ultrasonic cleaner to wash your glassware in organic chemistry experiments, saving time and convenience, then rinse them with water.

In addition to the above, vessels used for refining or organic analysis must be rinsed with distilled water. However, using various chemical reagents and organic solvents for cleaning instruments must be discouraged clearly because the use of various chemical reagents and organic solvents is wasteful and potentially dangerous.

### 1.6.2 Drying of common instruments

In addition to cleaning, glass instruments used for organic chemistry experiments often need to be dried. In some experiments, the apparatus is dry or not is the key point to the success or failure of the experiment. Generally, the cleaned apparatus can be used if there are no traces of water. However, some experiments must be strictly water-free; otherwise, the reaction is prevented from proceeding correctly. There are several ways to dry glass instruments:

(a) Air-drying

Air-drying means putting the cleaned instruments on a drying rack to dry, which is a common and simple method. However, care must be taken: if the glass instruments are not clean enough, water droplets will not flow down easily, drying will be slower, and stains will remain after drying.

(b) Blow-drying

Dry the instrument with an air dryer or a hair dryer.

(c) Drying with organic solvents

When in a hurry, organic solvents can aid drying by adding a small amount of ethanol or ether, then rotating the instrument to allow the solvent to flow through the inner wall, pouring it out, and then blowing it dry with a hair dryer to achieve rapid drying.

(d) High-temperature drying

Put the glass instrument into the oven for drying, with the instrument opening upward. For instruments with frosted glass plugs, the piston must be removed and dried. Keep the temperature in the oven at 100-155℃ for a while. It is best to take out the dried glass instruments after they are cooled to room temperature in the oven. Do not let the hot glass instruments get cold water to avoid breakage.

### 1.6.3 Maintenance methods of common instruments

The properties of various glass instruments in organic chemistry experiments are different. Their performance, maintenance, and washing methods must be mastered to use them correctly, improve experimental results, and avoid unnecessary losses. The following describes the maintenance and cleaning methods of several commonly used glass instruments.

(a) Separation funnel

The piston and top stopper of the separatory funnel are grinding, and if they are not original, they may not be tight. Therefore, it is important to protect them when using them and not interchange them. After use, make sure to put a piece of paper between the ground mouth of the piston and the top plug to prevent them from being difficult to open after a long time.

(b) Condenser

Condensate tubes are heavy when filled with water, so the clamps should be clamped to the center of gravity of the condensate tube to avoid tipping. If the inner and outer tubes are made of glass, they are unsuitable for use under high temperature conditions.

When washing the condensate tubes, use a long brush. If washing with organic solvents, the cleaned condensate tubes should be placed on a drying rack to dry for use.

(c) Distillation flask

The distillation flask's branch pipe is easily broken, so pay attention to protect the branch pipe of the distillation flask when using or in storage. The welding part of the branch pipe cannot be heated directly.

(d) Thermometer

The glass at the mercury bulb area of the thermometer is fragile and easily broken, so be careful when using it. The thermometer should not be used as a stirring rod, nor should it be used to measure temperatures above the maximum scale of the thermometer, nor should it be left in hot solvents for a long time; otherwise, the mercury bulb will be deformed, resulting in inaccurate temperature measurements.

The thermometer should be cooled slowly after use. Especially after measuring high temperatures, do not rinse with water immediately; otherwise, it will break or the mercury column will fracture open. After use, the thermometer should be hung on a metal stand, then washed and dried after cooling, and put back into the thermometer box, which should be padded with a small piece of cotton. If the thermometer is in a cardboard box, check that the bottom of the box is intact when putting it back in.

## 1.7 Application of low-temperature refrigeration

With the development of science and technology, refrigeration technology is also improving. Using deep cooling, many reactions that can not be carried out at room temperature, such as negative ion reactions or reactions with organometallic compounds, can be carried out successfully. Low-temperature operation is also commonly used in general organic experiments, such as diazotization reactions and nitrosation reactions. Some reactions do not require low temperature, but refrigeration is needed to transfer excess heat so that the reaction is carried out properly, so refrigeration technology plays an important role in the progress of organic chemistry.

Depending on the requirements of the reaction, different refrigerants can be used. Water can cool the reactants to room temperature; ice or ice-water mixture can cool the reactants to as low as 0℃; refrigerants obtained by mixing crushed ice with inorganic salts in the right proportions, the cooling temperature varies with the proportion of inorganic salts mixed and can reach temperatures of around 0 to −40℃; dry ice (solid carbon dioxide) or liquid ammonia mixed with organic solvents can give temperatures of below −70℃. More intensive cooling can be achieved using liquid nitrogen (up to −195.8℃). If the products need to be stored at a low temperature for a long time, the bottles containing the products should be labeled, tightly corked, and stored in a cryogenic refrigerator or freezer. When using cryogenic refrigerants, do not touch them directly with your hands and pay attention to avoid frostbite. When measuring low temperatures below −38℃, a mercury thermometer can not be used (the freezing point of mercury is −38.87℃), and a paraffin thermometer should be used.

## 1.8 Commonly heating appliances and equipment

### 1.8.1 Gaslamp

Gaslamp is one of the most common heating tools in the laboratory, mostly for heating aqueous and high boiling solutions. When using a gas lamp to heat a flask or other apparatus, it must be padded with asbestos mesh. When in use, the flame of the gas lamp can vary as the amount of air is adjusted. When the appropriate amount of air is introduced, the flame is made up of three parts: the inner flame (A) – a green cone

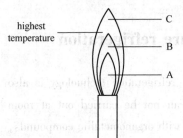

Fig 1-12 Schematic diagram of gaslamp flame

shape; the middle flame (B)-a dark blue; the outer flame (C)-a light blue, light blue and dark blue parts are high-temperature areas. The schematic diagram of gaslamp flame is shown in Fig 1-12.

### 1.8.2　Water bath

To eliminate the disadvantages of direct heating by open flame, various heating baths can be used depending on the experimental requirements. Water baths are one of the most commonly used. In use, the heated apparatus is immersed in water and heated to the desired temperature. Sometimes (as evaporating concentrated solutions), instead of immersing the vessel (beaker, evaporating dish, etc.) in water, it is placed on the lid of a water bath and heated by contact with water vapor. Both can heat liquids to around 95℃.

### 1.8.3　Oil bath

The oil bath is also a commonly used heating method, and the oil used is mostly linseed oil, castor oil, glycerine, silicone oil. The heating temperature is usually 100 to 250℃. When heating a flask, the flask must be immersed in oil. The disadvantages of an oil bath are: oily smoke will come out when the temperature rises, it can spontaneously combust when it reaches the ignition point, and an open flame can also cause a fire, and it is easy to be aged, become sticky and black after a long time. In order to overcome these disadvantages, silicone oil can be used. Silicone oil, also known as organosilicone oil, is a class of linear structured oily substance obtained by hydrolytic condensation of silicone monomers, generally colorless, odourless, non-toxic, non-volatile liquid, stable in nature, but more expensive.

### 1.8.4　Hair dryer

The hair dryer used in the laboratory can blow cold and hot air for drying glass instruments. It should be placed in a dry place to prevent moisture and corrosion.

### 1.8.5　Electric heating mantle

It consists of glass fibers wrapped with electric heating wires. When heating and distilling flammable organic materials, it has the advantage of not easily causing a fire and high thermal efficiency. A regulating transformer can control the temperature, and the maximum heating temperature can reach about 400℃. The volume of the electric mantle should generally match the volume of the flask. When it is used for distillation, the material in the flask gradually decreases, which then causes the flask wall to overheat, resulting in the distillate being scorched and affecting the distillation results. This is the reason why you should control the temperature when using it.

### 1.8.6 Rotary evaporator

The rotary evaporator comprises a motor-driven rotatable evaporator (round-bottom flask), condenser and receiver, which can be operated under atmospheric or reduced pressure. The evaporating liquid can be added at one time or in batches. Due to the continuous rotation of the evaporator, it can be exempted from adding zeolite, and no flash boiling will occur. When the evaporator rotates, it will make the evaporation surface of the liquid increase significantly and accelerate the evaporation rate. Therefore, it is a convenient device for concentrating solutions and recovering solvents.

### 1.8.7 Electric stirrer

Electric stirrers are generally suitable for use in solutions such as oil and water or in solid-to-liquid reactions. They are not suitable for overly viscous gelatinous solutions. It must be connected to the earth when in use. The stirrer should be kept clean and dry and protected against moisture and corrosion. The bearings should be lubricated with oil regularly.

### 1.8.8 Voltage regulating transformer

The voltage regulating transformer is a device that regulates the voltage of the power supply and is often used to regulate the temperature of the electric heating mantle, to regulate the speed of the electric stirrer. Notes when using:

( ⅰ ) The wires should be connected correctly and never misconnected, while the transformer should be connected to the earth.

( ⅱ ) The knob should be adjusted evenly and slowly to prevent sparks and damage to the contact points caused by violent abrasion.

( ⅲ ) After use, set the knob back to zero, cut off the power supply, and place it in a dry and ventilated place.

## 1.8.9  Magnetic stirrer

A magnetic stirrer is commonly used in experiments, consisting of a rotatable magnet and a potentiometer to controls the speed. When in use, the Teflon stirrer is placed into the reaction vessel and a suitable size stirrer can be selected according to the size of the vessel to achieve optimum stirring. The operating steps are as follows:

( i ) Turn on the power switch, and the indicator light is on.

( ii ) Turn on the stirring switch, the indicator light is on, and turn the knob to make the stirring bar rotate from slow to fast in the container.

( iii ) When control of the temperature is required, a thermocouple can be used to monitor the temperature and automatically stop the heating when the temperature rises to a pre-set temperature.

( iv ) Thermocouples are not necessary if automatic thermostats are not required for stirring and heating. It is important that the two leads do not touch each other during use.

( v ) Stirring is carried out at room temperature, and the thermocouple can be unplugged.

( vi ) To stop stirring, turn the knob slowly to zero, and then cut off the power.

The general performance of the magnetic stirrer used in the laboratory is:

( i ) Stirring speed: 0–1200r/min.

( ii ) Temperature range: room temperature to 100℃.

( iii ) Mixing capacity: 20–3000mL.

( iv ) Power: 300W, can work continuously.

Organic solvents, strong acids, alkalis, and other corrosive chemicals are strictly forbidden to leach the stirrer during use. After use, wipe the mixer clean and store it in a dry place.

# Chapter 2

# Basic Operation and Experimental Technology of Organic Chemistry Experiment

## 2.1 Determination of physical constants of organic compounds

### 2.1.1 Determination of melting point

It is generally accepted that when a solid compound is heated to a certain temperature, the solid melts and transforms into a liquid. The temperature at this point is the melting point of the compound. Strictly speaking, the melting point of a solid compound is the temperature at which the solid and liquid phases reach equilibrium at one atmosphere pressure. Most crystalline organic compounds have a fixed melting point, and most are below 300℃, which is easy to determine. However, there is a temperature interval between the start of melting and the complete melting of an organic compound in practical measurement experiments, and this temperature interval is called the melting range, also called the melting distance or melting point range. The melting range of pure compounds generally does not exceed 0.5℃. When a compound contains impurities, it tends to have a lower melting point and a longer melting range than the pure substance. Therefore, the melting point can be measured to estimate the purity of a compound.

(a) Basic principles

If the melting point of a compound is to be determined accurately, it must be determined with the aid of a phase diagram. When the solid and liquid phases of a compound are placed in the same container at a certain temperature and pressure, three situations may occur: the solid phase rapidly transforms into a liquid phase; the liquid phase rapidly transforms into a solid phase; the solid and liquid phases coexist (the temperature at which the solid and liquid coexist is the melting point). The proportion of solid and liquid phases at a given temperature can be judged from the vapor pressure of the compound as a function of temperature (phase diagram). The proportion of the phase with the higher vapor pressure is high relative to the phase with the lower vapor pressure. Fig 2-1(A) shows the relationship between vapor pressure and temperature for the solid phase, and Fig 2-1(B) shows the relationship between vapor pressure and temperature for the liquid phase. By superimposing the two curves in Fig 2-1(A) and (B), the curve in Fig 2-1(C) is obtained. The two curves intersect at point $M$, where the solid and liquid phases can coexist. The temperature $T$ at this point is the melting point of the compound, which is the reason why pure compounds have a fixed and sharp melting point.

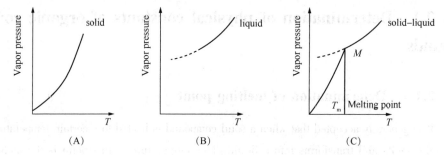

Fig 2-1　(A) the relationship between vapor pressure and temperature for the solid phase, (B) relationship between vapor pressure and temperature for the liquid phase, (C) superimposing the two curves in (A) and (B)

In most cases, the melting point of an organic compound is not determined by a phase diagram, and in practice is mostly determined by the capillary method. The melting point we observed is the temperature at which the solid starts to melt until it is completely melted, that is, the melting range, which is regarded as the compound's melting point.

(b) Micro melting point method

Micro melting point determination is a method of observing the melting process of a sample with the aid of a microscope. There are many different types of microscopic melting point measuring instruments and different instrumentation.

The microscopic melting point tester consists mainly of an electric heating system, a temperature measuring system, and a microscope. To determine the melting point, the sample is placed between two clean slides, placed in a heat bath, and the microscope is adjusted to observe the crystalline shape of the substance being measured. To determine the melting point, first turn on the power and adjust the heating knob so that the temperature rises quickly. When the temperature is 10-15℃ below the melting point, adjust the heating knob to slow down the increasing rate of temperature so that the temperature rises by 1 to 2℃ per minute.

When a repeat measurement is required, the heating is stopped and the metal heat sink is then placed in the heat bath. The heat sink serves to bring the temperature of the heat bath down very quickly.

Detail determination steps: the sample is finely ground and placed on a slide, taking care not to pile it up so that the shape of the sample can be observed in the eyepiece of the microscope. Turn on the heater and control the heating rate with a voltage regulating transformer. When the temperature is close to the sample's melting point, the heating rate is controlled at 1.0 to 1.5℃ per minute. When the crystalline angles of the sample become rounded and melting begins, the temperature is recorded as the initial melting temperature; just after the crystals have completely disappeared and the sample melts completely, the temperature is recorded as the full melting temperature. The advantage of this method is that it allows careful observation and the determination of high melting point compounds. Features of the micro melting point tester:

( ⅰ ) The microscope has a large field of view, a large working distance, and a strong stereo sense. Not easily fatigued by long hours of operation.

( ⅱ ) Intelligent temperature control with digital display of set and measured temperatures.

( ⅲ ) Intelligent regulation of the hot table temperature, with small flushing temperature, fast heating, automatic constant temperature.

### 2.1.2　Relative density and its determination

Relative density is an important constant for identifying liquid compounds and is the mass of a substance contained in a unit volume. It is usually expressed by $d_4^{20}$, meaning the ratio of the mass of the substance at 20℃ to the mass of the same volume of water at 4℃ (the density for water is 1.000g/cm$^3$ at 4℃). When g/cm$^3$ is used as the unit, the value of $d_4^{20}$ is the density of the substance, expressed by $\rho$. The density of a substance is related to its conditions (temperature, pressure); for solid or liquid substances, the effect of pressure on density is negligible. There are many ways to measure density, but here only describe the specific gravity bottle method.

A clean, dry pycnometer (1–5mL) is accurately weighed to 0.001g, and its mass is $m_0$, then the bottle is filled with a known density liquid. The capillary stopper is pushed slightly to the appropriate position (at this time, care should be taken not to have air bubbles in the bottle) and placed in a constant temperature bath; after 10min, the bottle is removed, the liquid level is adjusted to the scale of the bottle, the outer wall of the bottle is dried, and $m_1$ is weighed as the known liquid in the determination of temperature (because the volume of the liquid is related to the temperature, must make the bottle in the thermostat tank at a constant temperature, deviation of 0.03℃) of the mass. Pour off the known liquid, dry the bottle, and put it into the liquid to be measured; keep it at a constant temperature for 10min and weigh its mass in the same way to obtain $m_2$ as the mass of the liquid to be measured at the measuring temperature. Each weighing must be averaged from two weighings. Calculate the relative density of the liquid to be measured at the temperature $t$ using the following formula:

$$d^t = \frac{m_2 - m_0}{m_1 - m_0} \times d_1^t$$

Where $d_1^t$ is the relative density of the known liquid (usually water). If you want to measure the relative density of a certain liquid at $d_4^{20}$, you need only to control the measuring temperature at 20℃ and use water as a known liquid. The density of water at 20℃ is 0.99823g/cm³.

The commonly used unit of measurement for density should be expressed as $\rho$ (g/cm³).

### 2.1.3 Boiling point determination

Due to molecular movement, liquid molecules tend to escape from the surface, and this tendency increases with temperature. That is, a liquid at a certain temperature, there is a certain vapor pressure in equilibrium with it, and this vapor pressure changes with the change of temperature. As the temperature rises, the vapor pressure also rises. When a certain temperature is reached, the vapor pressure of the liquid is the same as the atmospheric pressure; at this point, the vapor inside the liquid is free to escape from the surface, and thus the phenomenon of boiling occurs. Therefore, when the vapor pressure of the liquid is equal to the standard atmospheric pressure, the temperature is called the boiling point of the liquid. For example, the boiling point of water is 100℃, which means that water boils at 100℃ under pressure of 101.3kPa (760mmHg).

The boiling point of a substance varies with the external atmospheric pressure. Therefore, when discussing or reporting the boiling point of a compound, it is important to indicate the external atmospheric pressure when the boiling point was determined, so that it can be compared with literature values.

The boiling point is one of the important physical constants of liquid organic compounds. Pure liquid organic compounds usually have a fixed boiling point, and the boiling point's determination allows a preliminary determination of the purity of the liquid.

The boiling point is determined by two methods: the constant method and the micro method. The apparatus for the constant method is the same as for the distillation operation. When the liquid is impure, the boiling range is very long (usually more than 3℃). In this case, it is not possible to determine the boiling point of the liquid, and the liquid should be purified by other methods before the boiling point is determined.

The boiling point can be measured by the micro methods using the apparatus shown in Fig 2 – 2. Add 1 or 2 drops of the liquid sample to the outer tube of the boiling point tube to a height of about 1 cm, then place the inner tube, fix the boiling point tube next to the thermometer, and then place it into the hot bath for heating. When heated, small bubbles slowly escape from the inner tube due to gas expansion, which will turn into a series of small bubbles escaping rapidly when the boiling point of the liquid is reached. At this point, the heating can

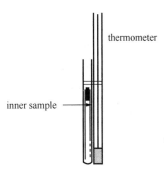

Fig 2-2  Schematic diagram of the microscope setup for boiling point measurement

be stopped, so that the temperature of the bath falls on its own, the speed of bubbles escaping gradually becomes slower. When the bubbles are no longer escaping and the liquid is just about to enter the inner tube (the last bubble is just about to retract to the inner tube), it means that the vapor pressure inside the capillary is equal to the external pressure, and the temperature at this point is the boiling point of the liquid. For calibration purposes, the temperature is lowered by a few degrees and then heated again, noting the temperature at the time when a large number of bubbles appear. The difference between the two thermometer readings should be no more than 1℃.

## 2.2 Simple glazier operation

In the process of conducting organic chemistry experiments, you often need to make some glass supplies yourself, such as glass bends for connecting some experimental devices at different angles, capillary tubes for melting point measurement, thin layer chromatography, reduced pressure distillation, as well as experiments often drops, stirring bars, glass rods, glass nails, and so on. A good knowledge of some simple glasswork techniques will bring great convenience to experimental work.

### 2.2.1 Cleaning and cutting of glass tube (rod)

Glass tubes (glass rods) should be cleaned and dried before processing. General organic chemistry experiments with glass tubes (glass rods) can be cleaned with water and detergent. If necessary, after cleaning can be rinsed with distilled water, and then dried in the oven.

The glass tubes (glass rods) can be cut to a certain length as required before processing. When cutting, use a file to file a slightly deeper mark in one direction at the place that needs to be cut; then hold the glass tube (glass rod) with two hands, hold the two sides of the back of the file mark with your thumb, slightly forward, while pulling to the left and right, you can make the glass tube (glass rod) neatly disconnected. Before fracturing, you can apply some water to the file mark, which is easier to break.

The cut glass tube (glass rod) is very sharp in section and must be melted in the fire to make the fracture smooth. When fusing, the glass tube (glass rod) can be burned at an angle of 45° on the edge of the oxide flame while turning back and forth until the section is smooth. A schematic diagram of the cutting of a glass tube (rod) is shown in Fig 2-3.

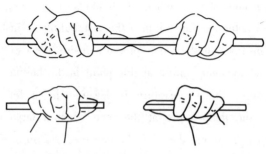

Fig 2-3　The cutting of glass tube (rod)

## 2.2.2 Drawing the capillary

The glass tube used for drawing capillary tubes differs from normal glass tubes in that it has a diameter of 1 cm and a wall thickness of approximately 1mm. The washed and dried glass tube is heated in a suitable position in a strong fire and is tilted at an angle to increase the area exposed to heat. The two-handed operation is essentially the same as for pulling a dropper. When the glass tube is soft, remove it from the flame, rotate it while keeping it horizontal, pull it away from the sides, place it flat on the table after it has cooled slightly, and pad the thick part of the tube at both ends with asbestos mesh. Once cooled, the capillary tube with a 1-1.5mm diameter is cut to the required length with a small grinding wheel.

## 2.2.3 Drawing dropper

The glass tube, 5-6mm in diameter and 15cm long, is first baked in a low flame at the stretching point to prevent it from bursting when exposed to strong heat, then the middle of the glass tube is heated over a strong flame, at which point one hand holds one end of the tube while the other hand holds the other end and turns it in one direction. When the glass tube starts to soften, the hand holding the glass tube should also rotate with the other hand at the same speed and direction to prevent the softened part from twisting. When the glass tube is yellow and soft, remove it from the flame and rotate it back and forth in the same direction while pulling it with both hands until it becomes the desired fineness. When the glass tube hardens, stop rotating, put on the asbestos net to cool, then use a small grinding wheel to cut off at the appropriate part of the thin tube and burn around the thin tube's edge at the mouth of the tube. The thick end of the glass tube is heated over high heat until it is yellow and soft, pressed vertically on the asbestos mesh so that the edges protrude, cooled, and then fitted with a latex tip, thus making two droppers. To draw the dropper, note that:

( i ) The glass tube should be constantly rotated when heated so that it is heated evenly. When the glass tube is yellow and soft, pay attention to the movement of both hands to avoid pulling the glass tube thin or twisting it when it is heated.

( ii ) Master the "heating temperature" of the glass tube melting, the "heating temperature" is not enough, and the drawn tube is too thick to meet the requirements.

( iii ) When pulling, do not force too hard, start a little slower, and then more quickly stretched. When pulling, both hands must be rotated back and forth in the same direction so that the center of the tube is symmetrical and the mouth of the tube is round.

### 2.2.4 Plug configuration and punching

Stoppers are an essential accessory for connecting and sealing instruments in the laboratory. Commonly used are cork and rubber stoppers. The rubber stopper will swell when exposed to organic solvents, but can seal the instrument. Cork plugs are flimsy and should be pressed tightly with a cork press before use to prevent the plug from cracking when punching or from inhaling more water during use.

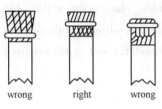

Fig 2-4 An example of a plug configuration

The size of the plug should be appropriate to the bore of the instrument used; the part of the plug entering the bore of the instrument should not be less than 1/3 of the height of the plug itself, nor more than 2/3. An example of a plug configuration is shown in Fig 2-4.

Sometimes, it is necessary to connect an instrument through a stopper or insert a glass tube or thermometer in a stopper. It is necessary to drill holes in the stopper. The size of the hole drilled should be such that the glass tube or thermometer can be inserted smoothly without any air leakage. Therefore, it is important to choose a hole punch of the correct diameter. The cork is loose in texture, and the hole punch should be slightly smaller than the outer diameter of the object to be inserted. When punching holes with rubber plugs, the diameter of the punch is approximately the same as the outer diameter of the inserted object. When drilling with a rubber stopper, the bore of the hole punch is approximately the same as the outer diameter of the inserted object. When drilling, place the cork on a board, hold the perforator, start from the small end of the cork vertical downward force evenly, and rotate in one direction, do not shake from side to side. When about 1/3 to 1/2 of the plug is punched in, the punch is pulled out by reversing the rotation, and the hole is punched through from the larger end of the plug to the original drilling position. When punching, you can apply glycerin or dip some water on the front end of the punch to reduce friction. Use even and slow force when punching; otherwise, the holes produced will be are small and unsuitable. If the hole is not rounded or is slightly smaller, use a round file to repair it. When the hole is finished, poke off the broken plugs in the perforator in time.

When inserting the glass tube or thermometer into the plughole, hold the tube or thermometer in your hand near the plug and screw it in slowly. The mouth of the tube may be dipped in a small amount of water or glycerine as a lubricant to reduce resistance. Remember that the hand holding the glass tube (or thermometer) should not be far from the stopper; otherwise, the glass tube (or thermometer) will easily break and cause a cutting accident.

The used stopper should be washed and dried for subsequent use. A contaminated stopper that cannot be cleaned should not be used again.

### 2.2.5 Bending of glass tube

Heat the cut glass tube on a low flame and then on a strong flame, turning it slowly while heating. When the glass is soft enough to bend, leave the flame, hold the tube horizontally in both hands and apply gentle pressure to the center so that it bends downwards under gravity. If you want to bend to a smaller angle, do this in several passes. First, bend into a large angle, then heating the glass tube, the heated part should be slightly offset; after several repeated operations, you can bend into the required angle. Each bend should not be too hard; otherwise, there will be deflated in the bend. Bend the glass tube should be in the same plane, the bending part of the arc. The bent glass tube should be on the same plane, and the bent part should be arc-shaped. When bending a glass tube, you can seal one end with a rubber tip and blow from the other end to keep the bend as thick or thin as it was. An example of a bending of glass tube is shown in Fig 2-5.

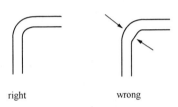

Fig 2-5  An example of a bending of glass tube

The bent glass tube should be immediately baked on a low fire for annealing and then placed on an asbestos mesh for cooling, which can prevent the glass tube from breaking due to the large internal stresses generated by rapid cooling.

## 2.3  Cooling and heating

### 2.3.1  Cooling

According to the low-temperature requirements of some experiments, the use of refrigerants is required for cooling operations in order to carry out reactions, separations, and purifications at low temperatures.

Coolant should be used in the following cases:

( i ) Some reactions whose intermediates are unstable at room temperature, should be carried out at a low temperature. For example, diazotization reactions are generally carried out at 0 to 5℃.

(ii) Exothermic reactions, which require cooling to control the rate of reaction.

(iii) To reduce the solubility of solids in the solvent in order to accelerate the precipitation of crystals.

(iv) Cooling of low-boiling point organics to reduce loss of reactants.

(v) High vacuum distillation device.

The choice of coolant should be based on the temperature required and the amount of heat absorbed during cooling:

(i) Water: water is cheap and has a high heat capacity, making it a common coolant.

(ii) Ice-water mixture: readily available coolant, the more crushed the ice, the better the effect.

(iii) Ice-salt mixture: salt or calcium chloride is added to crushed ice and can be chilled to -5 to -18℃.

(iv) Dry ice: can be chilled to below -60℃ and to -78℃ when added to a suitable solvent such as methanol or acetone.

(v) Liquid nitrogen: can be chilled to -196℃.

Note: when the temperature is below -38℃, a mercury thermometer cannot be used, and a paraffin thermometer is required.

## 2.3.2 Heating

Some organic reactions are challenging to occur or proceed slowly at room temperature. To increase the rate of the reaction, it is necessary to perform the reaction under heating conditions. In general, the rate of organic reaction accelerates with increasing temperature. Roughly every 10℃ increase in reaction temperature doubles the rate of reaction. In addition, many of the basic operations of organic chemistry experiments, such as reflux, distillation, dissolution, recrystallization, and melting, require heating.

The heaters commonly used in the laboratory include gas lamps, alcohol lamps, electric heaters, and enclosed electric furnaces. However, glass instruments are generally rarely heated directly by flames, because severe temperature changes and uneven heating will cause damage to the glass instruments. At the same time, due to local overheating, it may also cause partial decomposition of organic compounds and even possible explosion accidents. Therefore, different indirect heating methods are used in experiments according to different conditions. The easiest way to heat is through asbestos nets, but the heating is not uniform. It can not be applied in operations such as vacuum distillation or reflux of low-boiling flammable substances. In order to ensure uniform heating and operational safety, a suitable hot bath is used for indirect heating.

(a) Water bath

When the required heating temperature is below 80℃, the reaction vessel can be heated by immersing it in a water bath to bring the temperature of the water bath up to the required temperature. The water bath heats evenly and the temperature is easily controlled, making it suitable for heating and refluxing substances with low boiling points. The water level in the bath should be slightly higher than the level of the liquid in the reaction vessel. Pay attention that the reaction vessel can not be touched the bottom of the water bath. Water baths are usually made of copper or aluminum. Beakers can also be used instead of water baths when heating a small amount of low boiling substances. The lid of the water bath consists of a set of concentric rings of decreasing diameter, which effectively reduces the evaporation of water.

(b) Air bath

The heat source is allowed to heat the air, which in turn conducts the heat to the reaction vessel. An air bath can heat any liquid with a boiling point above 80℃. Direct heating of the vessel using a gas lamp through an asbestos net is the simplest air bath. However, it is not uniformly heated and unsuitable for low boiling point flammable liquids or reduced pressure distillation. An electric heating mantle is a better air bath, and can be heated from room temperature to about 200℃. When installing the heating mantle, keep the outer wall of the reaction flask about 2cm away from the inner wall of the heating mantle to prevent overheating. To facilitate temperature control, a voltage-regulating transformer can be connected.

(c) Oil bath

When the heating temperature is in the range of 100 to 250℃, an oil bath is more appropriate. The advantage of oil bath heating is that the reactants are heated evenly. The maximum temperature that can be reached in an oil bath depends on the type of oil used. In general, the temperature of the reactants should be about 20℃ lower than the temperature of the oil bath. Commonly used oil baths are paraffin oil (liquid paraffin), paraffin, glycerin, vegetable oil, and silicone oil.

( i ) Paraffin oil can be heated to around 200℃, does not decompose at higher temperatures, but is more combustible.

( ii ) Paraffin wax can be heated to about 200℃ and condenses to a solid when cooled to room temperature, making it easy to store.

(ⅲ) Glycerin can be heated to 140-150℃ and will decompose at a high temperature.

(ⅳ) Vegetable oils: rapeseed, castor, and soybean oils are available and heated to 220℃. Antioxidants such as hydroquinone (1%) are added for long-term use. It will decompose when the temperature is too high and may burn when the flashpoint is reached, so take care when using it.

(ⅴ) Silicone oil: stable at 250℃, good transparency, but more expensive. When heating with an oil bath, a thermometer should be placed in the oil bath (the thermometer should not touch the bottom of the bath) so that the temperature can be observed and adjusted at any time. In addition, do not splash water into the oil bath to avoid splashing.

(d) Sand bath

A sand bath should be used when the heating temperature is between 250 and 350℃. Usually, clean and dry fine sand is packed in an iron pan, the reaction vessel is half-buried in the sand, and the iron pan is heated. Keep a layer of sand at the bottom to prevent local overheating. However, due to the uneven temperature distribution of the sand bath, slow heat transfer, slow temperature rise, and too fast heat dissipation, the use range is limited.

(e) Other heating methods

In addition to the several common heating methods described above, other heating methods such as molten salt bath, metal bath (alloy bath), electrothermal method can be used. They should be selected according to experimental requirements and experimental conditions.

## 2.4 Separation and purification of organic compounds

### 2.4.1 Extraction

Extraction is one of the operations commonly used for the separation and purification of organic compounds. Extraction can be applied to extract the desired substance from a solid or liquid mixture, or it can be used to remove small amounts of impurities from a mixture.

(a) Basic principles

Extraction is an operation to separate or purify a substance according to the difference in solubility or distribution ratio between two solvents that are not mutually soluble. For example: a solution is made of organic substance X dissolved in solvent A. To extract X from it, we can choose a solvent B which has excellent solubility for X and does not react with solvent A and is not miscible, transfer the solution to a separatory funnel, add solvent B and shake it thoroughly. After standing, as solvent A and solvent B are not mutually soluble, they are divided into two layers. At this point, the concentration ratio of X between the two liquid phases of A and B is a constant at a certain temperature, which is called the "partition coefficient". This relationship is called the law of distribution. Available formulas:

$$K = \frac{\text{The concentration of X in solvent A}}{\text{The concentration of X in solvent B}}$$

When extracting with a given amount of solvent, is it better to do the extraction one time or multiple times? This can be illustrated using the following derivation: $V_1$ is the volume of the solution being extracted; $W_0$ is the total amount of (X) dissolved in the solution being extracted; $V_B$ is the volume of solvent B used each time. Obviously, after several extractions, the remaining amount of $W_n$ should be:

$$W_n = W_0 \left( \frac{KV_1}{KV_1 + V_B} \right)^n$$

When extracting with a certain amount of solvent, as $KV_1/(KV_1+V_B)$ in the above equation is constantly less than 1, the larger $n$ is, the smaller $W_n$ is. In other words, it is better to divide the solvent into several parts for several extractions than to use all the solvent for one extraction. For example, a solution containing 4g of n-butyric acid in 100mL of water is extracted at 15℃ if 100mL of benzene is used, and the distribution coefficient of n-butyric acid in water and benzene is $K = 1/3$. If 100mL of benzene is used for one extraction, the remaining amount of n-butyric acid in water after extraction is 1g, and the extraction efficiency is 75%; if 100mL of benzene is divided into three extractions, each with 33.3mL, the remaining amount is 0.5g, and the extraction yield is 87.5%. Therefore, the same volume of solvent in several extractions gives a higher yield than one extraction.

(b) Experimental operation

ⅰ) Extraction of substances in solution

The most application in experiments is the extraction of substances from aqueous solutions. A separatory funnel is commonly used in the laboratory to perform this operation. The size of the funnel should be more than twice as large as the volume of the liquid to be extracted. Clean the piston first and apply a layer of grease evenly, taking care not to block the piston hole. After plugging, rotate it a few times to distribute the grease evenly. Generally, you should check for leaks by adding water and make sure there are no leaks before using.

The schematic diagram of the extraction experiment operation of substances in solution is shown in Fig 2-6. Place the funnel in the iron ring, close the piston, pour the solution to be extracted and the extractant into the funnel from the upper mouth, and corking the upper mouth tightly (without greasing). Take care to tagger the slit in the stopper with the hole in the funnel neck. Hold the funnel stopper with the palm of your right hand, then hold the funnel with your thumb, index finger and middle finger, hold the funnel piston with your left hand, press the piston with your thumb, and shake the funnel up and down, tilting the top of the funnel downwards after every few shakes, with the lower branch pointing upwards at an angle (towards no one), open the piston with your thumb and index finger and release the gas generated by the shaking to equalise the internal and external pressure. After repeating the operation 2 to 3 times, the funnel is shaken vigorously for 2 to 3 minutes to bring the two immiscible liquids into full contact and improve the yield of the extraction. The funnel is then placed back into the iron ring and left to stand, aligning the hole in the upper mouth with the slit in the plug so that the pressure inside and outside is the same. When the two layers of liquid are separated, slowly open the piston and release the lower layer of liquid from the piston, then pour the upper layer of liquid out of the upper mouth of the funnel (not from the piston to avoid staining by the lower layer of liquid remaining in the branch tube). The aqueous layer is poured back into the funnel and extracted with a new extractant. The number of extractions depends on the partition coefficient. Generally, 3 to 5 extractions are sufficient. All extracts are combined and, if necessary, dried with a drying agent. The solvent is then evaporated, and the product obtained from the extraction is further purified by distillation or recrystallization, depending on its nature.

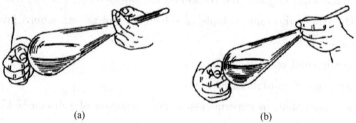

(a)                                      (b)

Fig 2-6  Schematic diagram of
the extraction experiment operation of substances in solution

In extraction, "salting out" reduces the solubility of organic matter in water or increases water's specific gravity and reduces emulsification.

ii ) Solid extraction

For the extraction of solid materials, the Soxhlet extractor is commonly used in the laboratory. The Soxhlet extractor uses the principle of solvent reflux and siphoning so that the solids are continuously extracted with pure solvent, and therefore the extraction yield is high. The solids are finely ground prior to extraction and then placed in a filter paper sleeve in an extractor which is connected to a flask containing solvent at the lower end and a condenser tube at the upper end.

As the solvent boils, the vapour rises through the glass branch pipe to the condenser, condenses into a liquid and then flows into the extractor. When the solution's level exceeds the siphon tube's highest point, a siphoning flow back into the flask occurs, thus extracting the part of the material dissolved in the solvent. After several repetitions of long-term reflux and siphoning, the solid soluble material is enriched in the flask. The solvent is then evaporated and the resulting extract is then purified using other methods. If the laboratory does not have a Soxhlet extractor, a constant pressure dropping funnel can also be used instead. The solid extraction experimental device diagram is shown in Fig 2-7.

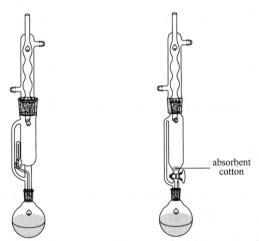

Fig 2-7  The solid extraction experimental device diagram

## 2.4.2  Recrystallization

Solid products obtained from organic reactions or from nature are often impure and may be interspersed with by-products, unreacted raw materials, solvents, and catalysts, and must be purified to obtain a pure product. A common and effective method for purifying solid organic compounds is recrystallization.

(a) Basic principles

The solubility of a solid organic compound in a solvent generally increases with temperature, so if a solid organic compound is dissolved in a hotter solvent to saturation, and then cooled to room temperature or below. The solution becomes supersaturated some of the crystals precipitate out. Using the difference in solubility of the solvent for the purified substance and the impurities, the purified substance is crystallized out of the supersaturated solution, leaving all or most of the impurities in the solution (or removed by filtration), thus achieving the purpose of purification.

(b) Operation steps

( i ) Select a suitable solvent.

( ii ) Dissolve the purified substance at the solvent's boiling temperature to make a near-saturated, concentrated solution.

( iii ) If the solution contains color impurities, boil the solution with an appropriate amount of activated carbon for decolorization.

( iv ) The boiling solution is filtered while still hot to remove insoluble impurities and activated carbon.

( v ) Cool the filtrate sufficiently to precipitate the crystals, leaving the soluble impurities in the mother liquor.

( vi ) Filtering under reduced pressure to separate the crystals from the mother liquor.

( vii ) Wash the crystals with a small amount of solvent to remove the adhering mother liquor.

( viii ) Dry crystals.

(c) Choice of solvent

The choice of a suitable solvent is the key to recrystallization and the ideal solvent should have the following conditions:

( i ) No chemical reaction with the substance to be purified.

( ii ) The solubility of the purified substance in the solvent should vary greatly with temperature, higher solubility at high temperature, and lower solubility at room temperature or lower.

( iii ) The solubility of impurities is either very large or very small (the former allows impurities to remain in the mother liquor and not precipitate out with the purified substance. The latter allows impurities to be removed during hot filtration).

(ⅳ) The boiling point is low, volatile, and easily separated from the crystals when dried.

(ⅴ) The purified substance forms better crystals.

(ⅵ) Non-toxic or very little toxicity, cheap, safe to handle, and easy to recover.

For some known compounds, information on solubility can be found in the chemical literature and then a suitable solvent can be selected. In many cases, however, the choice is made by experimental methods. When choosing a solvent, the principle of "similarity of solubility" should be taken into account.

The method is as follows: take 0.1g of the sample to be recrystallized in a small test tube, add the solvent drop by drop and keep shaking. If the solid is dissolved completely during the amount of solvent up to 1mL, the solvent is not suitable; if not dissolved completely, then heat carefully to boiling; if it is still not dissolved, continue to heat and add solvent in batches (0.5mL/time) up to 4mL. If the solid is still not dissolved completely under boiling, this solvent is also not suitable. Conversely, if the sample can be dissolved in 1 to 4mL of boiling solvent, cool the tube to room temperature or below and observe the crystallization. If the crystals can not be precipitated, rub the wall of the tube under the liquid surface with a glass rod to encourage the crystals to precipitate; if no crystals have precipitated, the solvent is not suitable. If the crystals can be precipitated normally, and the amount of crystals is also more, the solvent is proved to be suitable. When several solvents are tested in the same way, the best one is chosen by comparing the yield of crystallization, the ease of handling, the toxicity of the solvent, and the price.

Some compounds are either too soluble or too insoluble in a single solvent, making it difficult to choose a suitable solvent. In this case, a suitable mixture of solvents can be chosen. The so-called mixed solvent is a solvent that mixes two solvents that are particularly soluble in the compound and two solvents that are particularly insoluble in each other in a certain ratio and have good solvency properties. The appropriate amount of sample is first dissolved in the soluble boiling solvent, if there are insoluble impurities, filtered away while hot; if the impurities are colored, boiled with an appropriate amount of activated carbon for decolorization and filtered while hot. Then add another insoluble solvent while hot until the solution becomes cloudy, then heat or add the soluble solvent drop by drop until the solution is just clear and transparent. Finally, the solution is cooled to room temperature and the crystals are precipitated. This gives the ratio of the two solvent mixtures. If the mixing ratio of the two solvents is known, they can also be mixed first and then recrystallized. Commonly used solvent mixtures are: ethanol-water, ether-methanol, acetic acid-water, ether-acetone, acetone-water, ether-petroleum ether, pyridine-water, benzene-petroleum ether.

(d) Experimental operation

ⅰ) Dissolve the sample

When water is chosen as the solvent, the sample can be heated in a beaker or conical flask to dissolve it, whereas with organic solvents, to avoid evaporation and combustion, the sample must be heated in a reflux unit and the solvent added during heating should be added from the upper end of the condenser. The amount of solvent should be considered from two perspectives: on the one hand, to reduce dissolution losses, excess solvent should be avoided as much as possible; on the other hand, few solvent can reduce the yield during hot filtration by lowering the temperature and causing too many crystals to precipitate on the filter paper due to solvent evaporation. Therefore, to make recrystallization to obtain a purer product and higher yield, the amount of solvent should be appropriate, and solvent excess of about 20% is generally appropriate (Note: do not add unnecessary excess solvent because of insoluble impurities in the recrystallized material). Select the appropriate heat bath for heating according to the boiling point and flammability of the solvent.

ⅱ) Decolorization

If the solution contains color impurities, an appropriate amount of activated carbon can be added to decolorize the solution. The amount of activated carbon is generally 1 to 5% of the crude product. Too much activated carbon will adsorb some of the purified material and cause losses. When adding activated carbon, remove the fire, wait for the solution to cool slightly before adding it, and stir or shake to prevent boiling. After the activated carbon has been added, continue heating and generally boil for 5 to 10 minutes. If the decolorization is not good for one time, the operation can be repeated. The decolorization effect of activated carbon is related to the polarity of the solution and the amount of impurities. Activated carbon has a better decolorization effect in aqueous solutions and polar organic solvents, but a poorer decolorization effect in non-polar solvents.

ⅲ) Hot filter

The hot filtration method is used to remove insoluble impurities and activated carbon. If there are no insoluble impurities and the solution is clarified, this step can be omitted. Hot filtration under reduced pressure is fast, but because the hot solvent tends to evaporate under reduced pressure, cooling and concentrating the solution to the point of causing premature crystallization, the method of hot filtration under reduced pressure is often rarely used. The schematic diagram of the thermal filtration experimental device is shown in Fig 2-8.

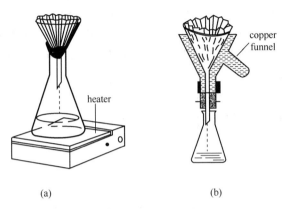

Fig 2-8　The schematic diagram of the thermal filtration experimental device

To avoid blocking the funnel's neck by cold crystals, use a short-necked or neckless glass funnel. Before filtering, place the funnel in an oven or under an infra-red lamp to warm it up before filtering, then remove the funnel and place it in an iron ring fixed on an iron stand or directly on a conical flask containing the filtrate. A folded fan of filter paper should be placed on top of the funnel at a height slightly higher than that of the funnel, with the outwardly protruding edges of the filter paper pressing against the funnel wall. When the above is ready, pour the boiling solution quickly into the filter paper, with the liquid level slightly below the top edge of the filter paper. If the solution cannot be poured at once, continue to heat the unfiltered solution over low heat to prevent it from cooling, but do not wait until the solution has been filtered completely before adding more. To reduce solvent evaporation, cover the funnel with an evaporating dish. If water is used as a solvent, the conical flask containing the filtrate can be heated over low heat to avoid crystallization on the filter paper due to the drop in temperature. However, when filtering volatile and flammable solvents, the fire source must be closed nearby, and the filter must not be heated. For substances that crystallize easily, or when the amount of solution to be filtered is large, a holding funnel can be used for filtration.

Fold the fan-shaped filter paper as shown in Fig 2-9. Fold the circular filter paper in half, and then fold it in four. Fold sides 2 and 3 to give side 4, 1 and 3 to give side 5, 2 and 5 to give side 6, 1 and 4 to give side 7, 2 and 4 to give side 8, 1 and 5 to give side 9, and 1 and 5 to give side 9. The shape of the folded filter paper is shown in Fig 2-9. Continue folding the filter paper in the opposite direction from one end to 1 and 9, 9 and 5, and up to 8 and 2 at the other end to make a fan shape. Open the double layer of filter paper and fold the same direction at 1 and 2 in opposite directions to obtain a fan-shaped folded filter paper interlaced inside and outside. Note: Do not fold the central round part of the filter paper too hard to avoid easy breakage during filtration. If your hands are not very clean when folding, the folded filter paper should be turned over gently before placing it in the funnel to avoid artificial impurities entering the solution.

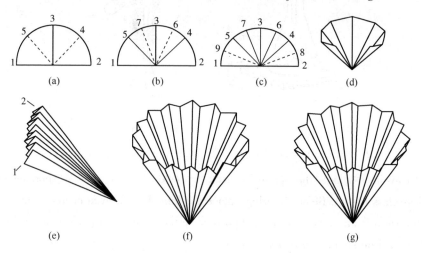

Fig 2-9 The schematic diagram of fold the fan-shaped filter paper

If the hot filtrate is left to cool slowly at room temperature, the crystals will precipitate slowly, resulting in larger, uniform, and pure crystals. Do not dip the filtrate in cold water to cool it quickly or shake the solution, as the crystals will not only be smaller, but the surface area will be large enough for the crystals to adsorb more impurities from the solution and affect the purity. However, the crystals must not be too large (more than 2mm), as they will intermingle with the solution and make it difficult to dry the crystals. If large crystals are forming, the crystals' average size can be reduced by shaking. If the crystals do not precipitate after cooling, rub the vessel's walls with a glass rod or put in crystalline seeds to cause the crystals to precipitate.

ⅳ) Suction filtration and washing of crystals

In order to separate the crystals from the mother liquor, the filter is usually pumped through a Buchner funnel. The pumping bottle is connected to the water pump by a more pressure-resistant rubber tube (it is best to connect a safety bottle between the two to avoid inadvertent back-siphoning of water from the pump into the pumping bottle). The schematic diagram of the suction filtration device is shown in Fig 2-10. The diameter of the circular filter paper in the filter funnel should be cut slightly smaller than the inner diameter of the funnel. Before pumping, wet the filter paper with a small amount of solvent and then turn on the pump so that the filter paper is sucked tightly to prevent the crystals from being sucked into the bottle from the gap at the edge of the filter paper when pumping. The crystals and the mother liquor are carefully poured into a Buchner funnel (also with the aid of a glass rod), the crystals remaining on the wall of the bottle can be rinsed several times with a small amount of filtrate and transferred to the Buchner funnel, the mother liquor is pumped out as far as possible and if necessary, a steel spatula can be used to squeeze the crystals in order to drain the mother liquor containing impurities adsorbed by the crystals. Then pull off the rubber tube attached to the stub of the filter bottle, or open the piston on the safety bottle and connect it to the atmosphere to avoid backflow of water.

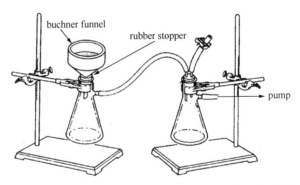

Fig 2-10  The schematic diagram of the suction filtration device

The mother liquor on the surface of the crystals can be washed with a solvent. A small amount of solvent (to minimize dissolution losses) is added to the crystals. The crystals are then gently tossed with a steel spatula to wet all the crystals and then drained again by connecting a vacuum pump. The washing is usually repeated 1 or 2 times to remove all the mother liquor from the crystal surface. When wetting the crystals with a small amount of solvent, disconnect the rubber tube from the filter bottle.

V) Drying of crystals

The crystals, which have been washed by filtration, have a small amount of solvent adsorbed on their surface and need to be dried by suitable drying methods to remove the solvent. The crystals are thoroughly dried before their melting point can be measured to check their purity.

Transfer the drained crystals to a clean petri dish with the aid of a steel spatula and spread them out. If the crystals do not absorb water, they can be left to dry naturally in the air (covered with a piece of filter paper or weighing paper to avoid dust staining); for thermally stable compounds, they can be dried in an oven at least 20℃ below the melting point of the compound or under an infrared lamp (care must be taken to control the temperature to prevent the crystals from melting). In the case of standard samples, analytical samples or samples prone to moisture absorption, the sample can be dried in a vacuum desiccator. Note: Crystals that sublimate easily at atmospheric pressure should not be dried under heat.

### 2.4.3 Distillation

Distillation is a common method for separating and purifying mixtures of liquids with different boiling points. It is also possible to determine the boiling point by distillation, which is significant for identifying liquid organic compounds.

(a) Basic principles

When the vapor pressure of the liquid increases to a level equal to the total pressure applied to the surface of the liquid (usually atmospheric pressure), a large number of bubbles escape from the inside of the liquid, this phenomenon is called boiling. The temperature at this point is called the boiling point of the liquid. The boiling point is related to the amount of external pressure. If the external pressure increases, the vapor pressure of the liquid boiling increases and the boiling point rises; on the contrary, if the external pressure is reduced, the vapor pressure of the boiling point decreases and the boiling point is lowered. In general, as an empirical rule, around 101.3kPa (760mmHg), when the pressure drops by 1.33kPa (10mmHg), the boiling point drops by about 0.5℃. At lower pressures, the boiling point of liquids drops by about 10℃ for every half of the pressure drop.

As the boiling point of a substance varies with the external atmospheric pressure. Therefore, when expressing the boiling point of a compound, it is essential to state the external atmospheric pressure. The boiling point is usually referred to as the boiling temperature at a pressure of 101.3kPa (760mmHg). For example, the boiling point of water is 100℃, which means that water boils at 100℃ under 101.3kPa pressure. The boiling point at other pressures should indicate the pressure. For example, at 12.3kPa (92.5mmHg), water boils at 50℃; at this time, the boiling point of water can be expressed as 50℃/12.3kPa. The combined operation of heating a liquid to boiling, turning the liquid into a vapour, and then condensing the evaporated vapour into a liquid is known as distillation. Obviously, distillation separates volatile and non-volatile substances, and also separates mixtures of liquids with different boiling points. However, to obtain a good separation, the boiling points of the liquid mixture components must be significantly different (at least 30℃ or more).

To avoid overheating during distillation and ensure a smooth boiling state, boiling aids such as zeolites, or capillary tubes closed at one end, are often added, effectively preventing bumping during heating. It should be noted that: never add a boiling aid to a liquid that is already close to or has already boiled! This is because if you add a boiling aid at this point, it will cause a bumping, and the liquid will easily rush out of the mouth of the bottle and even cause a fire. If you find that you have forgotten to add the boiling aid after heating, you should allow the liquid to cool down to below the boiling point before adding it. If distillation is stopped in the middle of the process, a new boiling aid should be added before reheating to avoid bumping.

(b) Process of distillation

The process of distillation can be divided into three stages as follows:

In the first stage, as the heating proceeds, the mixture in the distillation flask continues to vaporize. When the saturated vapor pressure of the mixture is equal to the atmospheric pressure, the liquid boils and begins to flow out as a result of the liquid being condensed. This part of the fraction is called the pre-fraction. Generally, the boiling point of this fraction is lower than that of the component to be collected and often is discarded as an impurity. Sometimes, the distilled liquid has almost no pre-fraction, but the first 1 or 2 drops of the distillation should be removed as a pre-fraction to rinse the apparatus to ensure the product's quality.

In the second stage, the temperature stabilizes in the boiling range, at which point the fraction is the desired product. The smaller the boiling range, the higher the purity of the components. As the fraction is distilled out, the volume of the mixed liquid in the distillation flask decreases until the temperature exceeds the boiling range and the reception can be stopped.

In the third stage, if only one component of the mixture needs to be collected, the remaining liquid in the distillation flask should then be discarded as a residue. In multi-component distillation, the temperature rises after the first component have been distilled and stabilizes at the boiling range of the second component. When the temperature stabilizes at the boiling range of the second component, the second component can be received. Regardless of the distillation operation, the liquid in the distillation flask must not be distilled dry to prevent the flask from overheating or exploding in the presence of peroxide.

When performing atmospheric distillation, the atmospheric pressure is usually not exactly equal to 101.3kPa (760mmHg) and therefore, strictly speaking, the thermometer should be calibrated. However, the deviation is usually small and therefore negligible.

All pure compounds have a fixed boiling point at a certain pressure, but liquids with a fixed boiling point are not necessarily pure compounds. When two or more substances form an azeotrope, they have the same composition in the liquid phase as they do in the gas phase. Therefore, they have the same composition at the same boiling point. Normal distillation methods cannot separate such a mixture.

(c) Experimental operation

ⅰ) Distillation unit and installation

A commonly used distillation apparatus is shown in Fig 2-11, consisting of a distillation flask, distillation head, thermometer sleeve, thermometer, straight condenser tube, elbow tube, and receiving flask.

The following points should be noted during the installation of the instruments:

( ⅰ ) The order of mounting the apparatus is bottom to up and left to right. The distillation flask should be placed at a distance of about 2cm from the bottom of the heating jacket to avoid local overheating. The apparatus should be set up straight so that it looks neat and tidy.

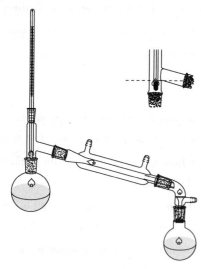

Fig 2-11 The schematic diagram of distillation apparatus

( ⅱ ) To ensure the accuracy of the measured temperature, the upper limit of the mercury bulb of the thermometer and the lower limit of the distillation head branch should be on the same level.

( ⅲ ) Condensate should flow in through the lower port of the condensate tube and out through the upper port to ensure that the casing of the condensate tube is always filled with water.

( ⅳ ) The distillation apparatus should not be closed up, as this may cause an explosion.

( Ⅴ ) Different condensate tubes should be used depending on the boiling point of the fractions. When the boiling point of the fraction is below 140℃, a straight condenser tube is generally used; when it is above 140℃, an air condenser tube is appropriate.

ii) Distillation operation

(i) Feeding: after the apparatus has been installed, remove the thermometer sleeve and thermometer, place a long-necked funnel on the distillation head and carefully pour the liquid to be distilled into the distillation flask, taking care not to allow the liquid to escape from the branch tube. Add several zeolites and check that the various parts of the apparatus are connected tightly and properly.

(ii) Heating: before heating, you should check whether the instrument is installed correctly, whether raw materials and zeolites are added, whether condensate is passed through, and only when everything is correct you can heat it. When the liquid boils and droplets appear on the mercury bulb of the thermometer, adjust the voltage appropriately so that there are condensed droplets on the mercury bulb of the thermometer. The distillation rate should be controlled at 1 to 2 drops per second.

(iii) Collect fractions and record the boiling range: before distillation, prepare two receiving flasks for collecting the pre-fraction and the fraction, respectively. Write down the thermometer reading at the start of the distillation and at the last drop, which is the boiling range (boiling point range) of the fraction. The liquid usually contains some high boiling point impurities. After the target fraction is distilled, the thermometer reading will rise significantly; if the original heating temperature is maintained, there will be no more distillate and the temperature will suddenly drop. At this point, the distillation should be stopped. Even if the content of impurities is minimal, do not distill it dry to avoid breaking the distillation bottle and other accidents.

(iv) Post-distillation: the power should be switched off to stop the heating and the voltage should be adjusted to zero. Then stop passing water and remove the apparatus. Remove the apparatus in the reverse order of installation, remove the receiving flask, remove the elbow tube, condenser tube, distillation head, distillation flask, and clean them.

## 2.4.4 Reduced pressure distillation

Reduced pressure distillation is an important method for separating and purifying organic compounds, suitable for high boiling point organic compounds or organic compounds prone to decomposition, oxidation, or polymerization when distilled at atmospheric pressure.

(a) Basic principles

The boiling point of a liquid varies with the external pressure. If the pressure of the system is reduced, the boiling point of the liquid is then reduced. Distillation at lower pressures is called reduced pressure distillation. The relationship between the boiling point of a substance and its pressure during reduced pressure distillation can be obtained in three ways:

( i ) Consult relevant manuals or dictionaries or reference books.

( ii ) The boiling point of a substance at different pressures can be deduced from an empirical approximation of the boiling point-pressure relationship graph. The boiling point-pressure relationship graph is shown in Fig 2-12. For example, the boiling point of ethyl acetoacetate at atmospheric pressure is 181℃. Now, to find how many degrees its boiling point is at 20mmHg. You should find the point at 181℃ on line $B$ and join this point at 20mmHg on line $C$ to form a straight line, extend this line to intersect with line $A$. At its intersection, the temperature shown at its intersection is the boiling point of ethyl acetoacetate at 20mmHg, which is about 82℃.

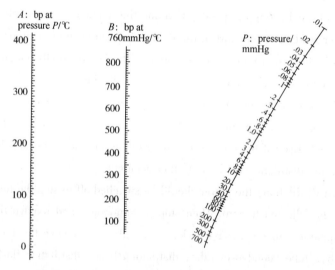

Fig 2-12　The boiling point-pressure relationship graph

( iii ) The boiling point at a given pressure can be calculated approximately from the following equation:

$$\lg P = A + B/T$$

$P$ is the vapor pressure, $T$ is the boiling point (absolute temperature), $A$ and $B$ are constants. If $\log P$ is the vertical coordinate and $1/T$ is the horizontal coordinate for the graph, a straight line can be approximated. The values of $A$ and $B$ can be calculated from the two sets of known pressures and temperatures. Then the boiling point of the liquid can be calculated by substituting the chosen pressure into the above equation.

(b) Vacuum distillation device

The former shown in the Fig 2-13 is a traditional reduced pressure distillation apparatus consisting of distillation, pumping (decompression), and a protection and pressure measuring device between them; the latter is a reduced pressure distillation apparatus commonly used now in laboratories (see Fig 2-14).

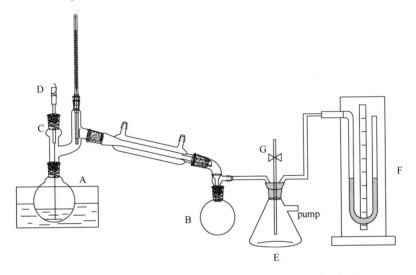

Fig 2-13　The schematic diagram of traditional reduced pressure distillation apparatus

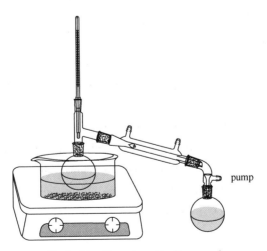

Fig 2-14　The schematic diagram of
reduced pressure distillation apparatus commonly used now in laboratories

ⅰ) Distillation part

The distillation flask A, used in the reduced-pressure distillation in the diagram, is a reduced-pressure distillation flask (also known as a Kerr distillation flask) with two necks. The purpose is to avoid the liquid inside the flask from boiling and rushing into the condensation tube during the reduced-pressure distillation. A thermometer is inserted into the mouth of the flask with a branch tube, and a capillary tube C is inserted into the other mouth, with the lower end of the capillary tube extending to about 1 to 2 mm from the bottom of the flask. The upper end of the capillary is connected to a piece of rubber tubing with a spiral clamp D. When distilling under reduced pressure, the spiral clamp is used to regulate the amount of air entering the liquid, a small amount of air into the liquid bubbles, becoming a liquid boiling gasification center, which can prevent the liquid from boiling, so that the boiling remains stable, which is very important for distillation under reduced pressure.

Nowadays, thanks to the use of a large number of electrical appliances in the laboratory, the operation of reduced pressure distillation has become simple and convenient. In reduced-pressure distillation, an electromagnetic stirring rotor is used as the vapourization center for the boiling liquid, making both the equipment and the operation easy and efficient. Of course, even though the operation has become more straightforward, the traditional precautions for reduced-pressure distillation must be observed.

ⅱ) Pumping part

Laboratories usually use pumps or oil pumps for depressurization. Nowadays, water circulation vacuum pumps are commonly used in laboratories for decompression. Water circulation vacuum pumps can also provide condensate, which is more convenient and practical in laboratories. The minimum pressure that the pump can achieve is the water vapor pressure at room temperature at the time. However, pumps are often not easy to obtain high vacuum levels due to their construction, water pressure, and water temperature. Oil pumps can reduce the pressure smoothly to 2-4mmHg and obtain a higher vacuum, but the structure of oil pumps is more precise. Therefore, when using the oil pump, you need to pay attention to protective maintenance, not to make the organic substances, water, acid, and other vapors into the pump. The oil can absorb the vapor of volatile organic substances in the pump, and the oil will be polluted, which will seriously reduce the efficiency of the oil pump; water vapor condensation in the pump will make the oil emulsified, which will also reduce the efficiency of the oil pump; and the inhalation of acidic vapor will corrode the oil pump.

ⅲ) Protection and pressure measuring device part

When using an oil pump for reduced pressure distillation, a buffer bottle, a cooling trap, a mercury manometer, a drying tower, and several absorption towers should be installed in sequence between the receiver and the oil pump. The purpose of the buffer bottle is to keep the pressure in the apparatus from changing too abruptly and prevent the pump oil from being sucked back. The cooling trap can be placed in a wide-mouthed thermos bottle with coolant. The choice of coolant depends on the need, such as ice-water, ice-salt, dry ice and acetone, etc. The purpose is to condense down the low boiling point organic solvent in the decompression system to protect the oil pump. The type of absorbent in the absorption tower is often based on the nature of the distillate, using anhydrous calcium chloride, solid sodium hydroxide, activated carbon, paraffin flakes, and molecular sieve, etc. The purpose is to absorb acidic gases, water vapor, and organic vapor. If a water circulation vacuum pump is used to reduce the pressure, no absorption device is required.

The pressure of a reduced pressure distillation can be measured with a mercury manometer. In open mercury manometers, the difference in height between the two arms of the mercury column is the difference between the atmospheric pressure and the pressure in the system, so the actual pressure (vacuum) in the distillation system should be the atmospheric pressure minus the pressure difference. In a closed mercury manometer, the difference in height between the two arms of the liquid is the vacuum in the distillation system. The schematic diagram of mercury manometer is shown in Fig 2-15.

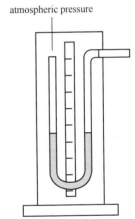

Fig 2-15　The schematic diagram of mercury manometer

(c) Vacuum distillation operation

Before starting the distillation, the gas-tightness of the apparatus must be checked. Fill the distillation flask with about 1/3 to 1/2 of its capacity of distilled material. Clamp the rubber tube on the capillary tube completely, open the safety bottle cork and turn on the vacuum pump. Gradually close the safety bottle stopper and observe the vacuum level achieved by the apparatus from the mercury manometer.

After checking, if the apparatus set-up is perfectly in order, distillation can begin. Calculate the vacuum level from the manometer reading and find out the boiling point of the liquid at that pressure. Turn on the condensate and choose a suitable heat bath to heat the distillation. When heating, at least 2/3 of the flask should be immersed in the liquid of the heat bath, but be careful not to let the bottom of the flask touch the bottom of the bath. Gradually increase the temperature of the heat bath, and the temperature of the bath should generally be about 20-30℃ higher than the boiling point of the distilled liquid, so that the liquid is kept boiling smoothly and the distillate flows out at a rate of 1 to 2 drops/second. During the distillation process, attention should be paid to the mercury manometer readings, recording data on time, including pressure, liquid boiling point, hot bath liquid temperature, and distillate outflow speed. If the boiling point of the starting distillate is lower than the boiling point of the desired fraction, replace the receiver when the desired temperature is reached.

When the distillation is completed, stop the heating, remove the hot bath, open the safety bottle cork slowly to bring the apparatus to atmospheric pressure. This operation must be careful and slowly unscrew the cork so that the mercury column in the manometer slowly returns to its original state. If the introduction of air too quickly, the mercury column will appear to break, and the mercury rises quickly in the closed mercury manometer. In that case, there is the possibility of breaking through the U-tube manometer, then turn off the vacuum pump. Do not disassemble the instrument until the pressure in the apparatus is equal to atmospheric pressure.

### 2.4.5 Steam distillation

Steam distillation is one of the common methods for separating and purifying liquid or solid organic compounds. Organic compounds that can be purified by steam distillation must have the following conditions: insoluble (or almost insoluble) in water; coexisting with water for a long time at about 100℃ without chemical changes; must have a certain vapor pressure at around 100℃ (generally not less than 1.33kPa).

A better separation can be obtained by steam distillation if one of the following conditions is present: organic compounds with a high boiling point, which can be separated from the by-products at atmospheric pressure but are easily destroyed; mixtures containing large amounts of resinous or tar-like substances which are difficult to separate by distillation or extraction; separation of adsorbed liquids from a larger number of solid reactants.

(a) Basic principles

When a substance that is immiscible with water is present together with water, the vapor pressure $P$ of the mixture, according to Dalton's law of partial pressure, should be the sum of the vapor pressure $P_A$ of the water and the vapor pressure $P_B$ of the substance, namely:

$$P = P_A + P_B$$

$P$ increases with temperature, and the mixture begins to boil when the temperature rises to the point where $P$ equals the external atmospheric pressure. At this point, the temperature is the mixture's boiling point, which must be lower than the boiling point of any of the mixture's components. Thus, when steam distillation is carried out by passing water vapor through an organic substance that is insoluble in water, the substance can be distilled out with water at a much lower temperature than that of the substance, and even lower than 100℃. What is distilled out is water and substances that are not miscible with water and can be easily separated, thus achieving purification.

In the distillate, the ratio of the mass of water to the mass of organic matter is:

$$\frac{m_A}{m_B} = \frac{M_A P_A}{M_B P_B}$$

Water has a low relative molecular mass and a large vapor pressure. Their product $M_A P_A$ is small, making it possible to separate substances with higher relative molecular masses and lower vapor pressures. For example, when water vapor is passed into a reaction mixture of aniline, which has a boiling point of 184.4℃. The mixture of aniline and water boils at 98.4℃; at this temperature, the vapor pressure of aniline is 5.73kPa, and the vapor pressure of water is 94.8kPa. The relative molecular masses of aniline and water are 93 and 18, respectively; and the ratio of the masses of aniline and water in the distillate is equal to:

$$93 \times 5.73 / 18 \times 94.8 = 1/3.3$$

Thus, the distillation of 3.3g of water would bring out 1g of aniline. Theoretically, the distillate should contain 23% aniline. However, in practice, the percentage obtained is lower because some of the water vapor leaves the distillation flask before it can come into full contact with the distilled material, and aniline is slightly soluble in water, so this calculation is only an approximation. Another example is the distillation of bromobenzene by steam distillation, which has a boiling point of is 135℃ and is not miscible with water. It starts to boil when heated to 95.5℃; at this time, the vapor pressure of water is 86.1kPa, and the vapor pressure of bromobenzene is 15.2kPa. From the calculation, the ratio of the mass of water and bromobenzene in the distillate is 6.5/1, and bromobenzene accounts for 61% in the distillate, and the distillate contains more bromobenzene than water. However, when a compound's molecular weight is large and its vapor pressure is too low, it cannot be purified by steam distillation, which requires a vapor pressure of at least 1.33kPa around 100℃. If the vapor pressure is between 0.13 and 0.67kPa, its content in the distillate is only 1%, or even lower. In order to increase the content in the distillate, it is necessary to find a way to increase the vapor pressure of the substance, which means raising the temperature so that the vapor temperature exceeds 100℃ and using superheated water vapor to distill, thus increasing the content of the substance in the distillate.

(b) Experimental operation

A commonly used steam distillation apparatus is shown in Fig 2-16 and consists of a steam generator, distillation section, condensing section, and receiver. A is the steam generator, and the side glass tube C is a liquid level gauge. You can observe the height of the liquid level in the generator. Usually, the amount of water is 3/4 of its volume. If it is too full, the water vapor will wash the water into the flask when boiling. The safety glass tube B should be inserted close to the bottom of generator A. When the water vapor pressure in the container is high, the water can rise along the glass tube to regulate the pressure inside the container. If water is ejected from the top of the glass tube, the whole system should be checked for blockages (usually the lower opening of the steam tube in the round-bottom flask is blocked by resinous or tar-like substances).

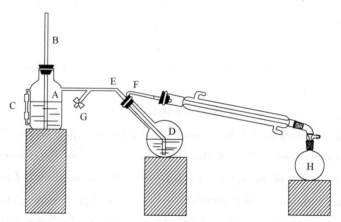

Fig 2-16  The schematic diagram of steam distillation apparatus

The distillation section is usually a 500mL long-necked round-bottom flask D, which should not contain more than 1/3 of its volume. To prevent the liquid in the flask from ricocheting into the condenser tube due to splashing, the flask is positioned at an angle of 45° towards the direction of the generator. The end of the vapor pouring tube E should be bent so that it faces the center of the flask vertically and is close to the bottom of the flask. The vapor pouring tube F (bent at an angle of approximately 30°) should preferably have a larger bore than tube E. One end is inserted into a double-bore cork with approximately 5mm exposed and the other end connected to the condenser tube. The distillate enters the receiver H through the elbow tube (the periphery of the receiver can be cooled in a cold water bath, if appropriate). A T-tube should be fitted between the water vapor generator and the long-necked round-bottomed flask. A spring clamp G is attached to the lower end of the T-tube in order to remove the condensed water droplets in time. The distance between the water vapor generator and the long-necked flask should be kept as short as possible to reduce the condensation of water vapor.

When performing steam distillation, place the mixture to be separated in round-bottom flask D, heat the steam generator (do not forget to add water to heating; otherwise, it will cause the solder to melt and damage the steam generator), until the water boils before clamping G, the water vapor is passed into D. In order that the water vapor does not accumulate too much in D due to condensation, if necessary, an asbestos net can be placed under D and heated over a low flame. Pay attention to adjusting the gas lamp to heat the steam generator, so that the water vapor is not produced too quickly, so as not to wash the mixture in D to the condenser tube, and so that all the steam can be condensed by the condensation tube. If the substance evaporating with the water vapor has a high melting point and is likely to precipitate solids after condensation. In that case, the flow rate of the condensate should be reduced so that it remains in a liquid state after condensation. If solids have been precipitated and close to blocking, the condensate flow can be temporarily stopped or even drained to allow the substance to melt and flow into the receiver with the water. Distillation can usually be stopped when the distillate is clear and no longer contains oil droplets of organic substances. When distillation needs to be interrupted or when distillation is completed, be sure to open the spring clamp G to bring the system into contact with the atmosphere before stopping the heating; otherwise, the liquid in D will be sucked back into A. In the distillation process, if you find that the water level in the safety tube B rises rapidly, you should also immediately open the spring clip, remove the heat source, and then continue with the steam distillation after the blockage has been removed. A Kjeldahl distillation flask can be used instead of a round-bottomed flask when distilling a small amount of material with steam distillation, or sometimes a three-necked flask can be used directly in place of a round-bottomed flask for the reaction.

( i ) Use a Kjeldahl distillation flask (head) for steam distillation of a small amount of material, and Kjeldahl distillation flask (head) for distillation apparatus is shown in Fig 2-17.

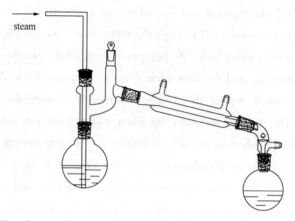

Fig 2-17  The schematic diagram of Kjeldahl distillation flask for distillation apparatus

( ii ) Use a three-necked flask instead of a round-bottomed flask for steam distillation is shown in Fig 2-18.

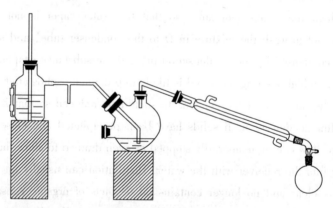

Fig 2-18  Schematic diagram of use a three-necked flask instead of a round-bottomed flask for steam distillation

(c) Superheated steam distillation device

Compounds with lower vapor pressures around 100℃ can be distilled using superheated steam. For example, a section of copper tube (preferably spiral) can be connected in series between a T-tube and a flask. The copper pipe is heated with a flame to increase the temperature of the steam. The flask is then kept warm with an oil bath. This can also be done with the following apparatus (the superheated steam distillation device is shown in Fig 2-19). A is used to remove condensed droplets from the steam, and B is a rigid glass tube wrapped in several layers of asbestos paper and heated with a flame. C is a thermometer sleeve with a thermometer inserted, and the flask is maintained at the same temperature as the steam in an oil or air bath.

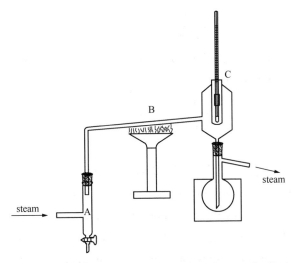

Fig 2-19 The schematic diagram of the superheated steam distillation device

## 2.4.6 Fractionation

A mixture of two or more mutually soluble liquids is difficult to separate by simple distillation if their boiling points are relatively close. In this case, a fractionation column can be used for separation, and this is fractional distillation. Fractional distillation is effectively the equivalent of multiple distillations. It is widely used in the laboratory and in the chemical industry to separate and purify mixtures.

(a) Basic principles

We will only discuss the case where the mixture is an ideal solution of binary components for simplicity. An ideal solution is one in which the components are mixed without thermal effects, without any change in volume, and which obeys Raoult's law. In this case, the vapor pressure of each component of the solution is equal to the product of the vapor pressure of the pure substance and its molar fraction in the solution. That is:

$$P_A = P_A^0 x_A^{liquid} \qquad P_B = P_B^0 x_B^{liquid}$$

$P_A^0$ and $P_B^0$ are the vapor pressure of pure A and B, respectively; and $x_A^{liquid}$ and $x_B^{liquid}$ are the mole fraction of A and B in the solution, respectively.

Total vapor pressure of the solution: $P = P_A + P_B$

According to Dalton's law of partial pressure, the vapor pressure of each component in the gas is proportional to its mole fraction. Therefore, in the gas phase the composition of the vapor of each component is:

$$x_A^{gas} = \frac{P_A}{P_A + P_B} \qquad x_B^{gas} = \frac{P_B}{P_A + P_B}$$

As deduced from the above equation, the relative concentrations of component B in the gas phase and the solution is:

$$\frac{x_A^{gas}}{x_B^{gas}} = \frac{P_B}{P_A + P_B} \times \frac{P_B^0}{P_B} = \frac{1}{x_B^{liquid} + x_A^{liquid} \dfrac{P_A^0}{P_B^0}}$$

Since in solution $x_A^{liquid} + x_B^{liquid} = 1$, if $P_A^0 = P_B^0$, then $x_B^{gas}/x_B^{liquid} = 1$, indicating that the compositions of the liquid phase and the gas phase are identical at this point, so that A and B can not be separated by distillation or fractionation. If $P_B^0 > P_A^0$, then $x_B^{gas}/x_B^{liquid} > 1$, indicating that the lower boiling point of B is more concentrated in the gas phase than in the liquid phase (a similar discussion can be made when $P_A^0 > P_B^0$). If the liquid is obtained by first distillation, the B component is more than the original liquid. If the resulting liquid is distilled for the second time, the volatile components will increase in the liquid after its vapor is condensed. After many repetitions, the two components can finally be separated (where the formation of an azeotropic mixture is not included in this example). Thus, fractional distillation is the use of fractionation columns to achieve this repeated distillation process.

A fractionation column consists mainly of a long, vertical, hollow tube with a certain shape to the column body, which is often filled with a special packing. The overall aim is to increase contact between the liquid and gas phases. When the boiling mixture enters the fractionation column, the condensate contains more substances with higher boiling points because the components with higher boiling points are easily condensed. In contrast, the vapor contains relatively more components with lower boiling points. When the condensate flows downwards and comes into contact with the rising vapor, heat is exchanged between them, meaning that the high boiling point substances in the rising vapor are condensed down and the low boiling point substances are still in vapor, while the low boiling point substances in the condensate are vaporized by heat and the high boiling point ones are still in liquid form. In this way, the liquid and gas phases are exchanged for many times, so that the low-boiling substances rise continuously and are finally distilled out, while the high-boiling substances flow back into the heated vessel, thus separating the substances with different boiling points.

(b) Simple fractionation device

The fractionation device is similar to the simple distillation device with the exception that a fractionation column is added between the distillation flask and the distillation head. The schematic diagram of the fractionation device is shown in Fig 2-20. There are many types of fractionation columns. Wechsler fractionation columns are commonly used in laboratories. Packed columns are generally used for semi-micro experiments; a glass tube is filled with inert materials, such as glass, ceramic, or metal chips in various shapes such as spiral and saddle shapes.

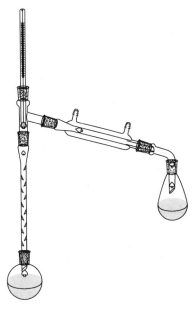

Fig 2-20  The schematic diagram of the fractionation device

(c) Simple fractionation operation

The simple fractionation operation is approximately the same as distillation. The mixture to be fractionated is placed in a round-bottomed flask, and zeolite is added; the column can be wrapped in asbestos cloth, which reduces the heat emission from the column and reduces the effect of airflow and room temperature; heating is carried out after careful inspection. After the liquid has boiled, care should be taken to adjust the temperature so that the vapor rises slowly into the fractionation column. When the vapor rises to the top of the column, droplets appear on the mercury bulb of the thermometer. Adjust the temperature of the bath so that the speed of the evaporated liquid is controlled at a drop every 2 to 3 seconds, so that a better fractionation effect can be obtained, and then gradually increase the temperature after the low boiling point component has been evaporated. When the second component is evaporated, a rapid rise in boiling point occurs. In this way, the liquid organic compounds are fractionated according to the boiling point of each component in turn.

The following points must be noted for good fractionation:

( i ) Fractional distillation should be carried out slowly and a constant distillation rate should be controlled.

( ii ) In general, maintain the temperature gradient within the fractionation column is achieved by adjusting the speed of the distillate. If the heating rate is fast, the evaporation rate is also fast, the temperature gradient within the column becomes smaller, affecting the separation effect; if the heating rate is too slow, the column will be blocked by the condensate, resulting in liquid flooding phenomenon. Therefore, a sufficient amount of liquid needs to flow from the fractionation column back to the flask; choose a suitable reflux ratio; the larger the reflux ratio, the better the separation effect.

(ⅲ) Choose the appropriate method to maintain a constant temperature inside the column. If necessary, wrap the fractionation column with a certain thickness of insulation material to ensure the temperature gradient inside the column.

## 2.5 Drying and the use of desiccant

Drying is a common method of removing small amounts of water or small amounts of organic solvents from solids, liquids, or gases. For example, many organic reactions need to be carried out under absolutely water-free conditions, the raw materials and solvents used should be dry; some compounds contain moisture and will deteriorate when heated, so they must also be dried when distilled or recrystallized; organic compounds need to be dried before qualitative or quantitative analysis and spectroscopy can give accurate results; some organic compounds may form azeotropic mixtures with small amounts of water or react with water to affect the product. Therefore, drying is one of the most common and important basic operations.

### 2.5.1 Basic principles

Drying methods of organic compounds can be divided into two types of methods: physical and chemical. Among the physical methods are: drying, air-drying, adsorption, freezing, fractionation, azeotropic distillation, and so on. In recent years, ion exchange resins and molecular sieves are also commonly used for drying.

The chemical method is the use of desiccant dehydration, which can be divided into two categories according to the dehydration effect.

Combine reversibly with water to form hydrates, such as:
$$CaCl_2 + 6H_2O \longrightarrow CaCl_2 \cdot 6H_2O$$

Irreversible chemical reaction with water to form new compounds, such as:
$$2Na + 2H_2O \longrightarrow 2NaOH + 2H_2 \uparrow$$

## 2.5.2 Drying of liquid organic compounds

(a) Selection of desiccant

For the drying of liquid organic compounds, the desiccant is usually used to come into direct contact with it; therefore, no chemical reaction occurs between the desiccant and the compound to be dried, including dissolution, coordination, conjoining, and catalysis. For example, an acidic substance cannot use an alkaline desiccant, while an alkaline substance cannot use an acidic desiccant.

When selecting a desiccant that combines with water to produce a hydrate, the desiccant's water absorption capacity and drying efficiency must be considered. The water absorption capacity refers to the amount of water absorbed by the desiccant per unit, and the drying efficiency refers to the degree to which the liquid is dried when it reaches equilibrium. For example, anhydrous sodium sulfate can form $Na_2SO_4 \cdot 10H_2O$, that is, 1g $Na_2SO_4$ can absorb up to 1.27g of water at most, and its water absorption capacity is 1.27. However, the water vapor pressure of its hydrate is also larger (255.98Pa at 25℃), so the drying efficiency is poor. Calcium chloride can form $CaCl_2 \cdot 6H_2O$, and its water absorption capacity is 0.97, and the water vapor pressure of this hydrate at 25℃ is 39.99Pa. Therefore, although the water absorption capacity of anhydrous calcium chloride is smaller, its drying efficiency is strong, so the suitable desiccant should be chosen according to the requirement of removing water when drying operation. Usually, this kind of desiccant needs a certain equilibrium time to form a hydrate, so it must be placed for a period of time after adding desiccant to achieve the effect of dehydration.

The desiccant which has absorbed water will be dehydrated again after being heated, and its vapor pressure will increase as the temperature rises. Therefore, the desiccant must be filtered out before the distillation of the dried liquid.

(b) The amount of desiccant

It is very important to master the dosage of desiccant. If the dosage is not enough, it is impossible to achieve the purpose of drying; if the dosage is too much, it will cause the loss of liquid due to the adsorption of desiccant. Take ether as an example, the solubility of water in ether is 1−1.5% at room temperature; if anhydrous calcium chloride is used to dry 100mL of ether containing water, all of it will turn into $CaCl_2 \cdot 6H_2O$, and its water absorption capacity is 0.97. That is to say, 1g of anhydrous calcium chloride can absorb about 0.97g of water, so that the theoretical dosage of anhydrous calcium chloride is less than 1g, but in fact, it is far more than 1g. There are still suspended microscopic water droplets in the ether layer, and the formation of high hydrates need a long time; it is often impossible to reach the proper water absorption capacity, so the actual amount of anhydrous calcium chloride input is greatly excessive, often need to use 7 to 10g of anhydrous calcium chloride. When operating, generally put a small amount of desiccant into the liquid and shake it. If there is desiccant adhering to the wall or bonding with each other, it means that the amount of desiccant is not enough and desiccant should be added again; if water phase appears after putting in desiccant, the water must be dried up with a pipette and then new desiccant should be added.

Before drying, the liquid is cloudy and becomes clarified after drying, which can simply be taken as a sign that the water has been largely removed. General, the general dosage of desiccant is about 0.5-1g per 10mL liquid. Due to the different water content, the difference in desiccant quality, the size of desiccant particles, and the different temperatures when drying, it is not easy to specify the certain quantity, and the above quantity is for reference only.

(c) Commonly used desiccants

We can see the commonly used desiccants in Table 2-1.

**Table 2-1 The Summary of commonly used dryers**

| Desiccant | Water absorption | Drying efficiency | Drying speed |
|---|---|---|---|
| $CaCl_2$ | Form $CaCl_2 \cdot nH_2O(n=1,2,4,5)$ | medium | Quick |
| $MgSO_4$ | Form $MgSO_4 \cdot nH_2O(n=1,2,4,5,6,7)$ | Weaker | Quick |
| $Na_2SO_4$ | Form $MgSO_4 \cdot 10H_2O$ | Weaker | Slow |
| $CaSO_4$ | Form $2CaSO_4 \cdot H_2O$ | Strong | Slow |
| CaO | Form $Ca(OH)_2$ | Strong | Quick |
| $P_2O_5$ | Form $H_3PO_4$ | Strong | Fast |
| Na | Form NaOH | Strong | Quick |
| Molecular sieve | Physical adsorption | Strong | Slow |

## 2.5.3 Drying of solid organic compounds

Solids from recrystallization often carry water or organic solvents and should be dried by an appropriate method depending on the nature of the compound.

(a) Air-drying

This is the simplest method of drying. The solid to be dried is first placed on filter paper in a porcelain funnel or pressed on top of the filter paper. Then spread it out thinly on top of one piece of filter paper, cover it with another piece of filter paper, and let it dry slowly in the air.

(b) Heat drying

For heat-stable solid compounds can be dried in an oven or under an infra-red lamp. The heating temperature should not exceed the melting point of the solid to avoid discoloration or decomposition of the solid; if required, then dry in a vacuum constant temperature oven.

(c) Desiccator drying

For solids that are prone to moisture absorption or decomposition or discoloration when dried at higher temperatures, desiccators can be used to dry them. There are ordinary desiccators, vacuum desiccators, and vacuum constant temperature desiccators.

## 2.5.4 Drying of gas

Gases commonly used in organic chemistry experiments are $N_2$, $O_2$, $H_2$, $Cl_2$, $NH_3$, $CO_2$, etc. Sometimes it is required that the gases contain little or almost no $CO_2$, $H_2O$, etc., and then the above gases need to be dried. Gases are often dried in drying tubes, drying towers, gas bottles, etc. The commonly used desiccants for drying gases are listed in the following Table 2-2.

Table 2-2 Common desiccants for gas drying

| Desiccant | Gas that can be dried |
| --- | --- |
| CaO, NaOH, KOH | $NH_3$ |
| Anhydrous $CaCl_2$ | $H_2$, HCl, $N_2$, $O_2$, $CO_2$, $SO_2$, alkanes, ethers, alkenes, halogenated hydrocarbons |
| $P_2O_5$ | $H_2$, $N_2$, $O_2$, $CO_2$, $SO_2$, alkanes, alkenes |
| Conc. $H_2SO_4$ | $H_2$, HCl, $N_2$, $O_2$, $CO_2$, $SO_2$ |

# Chapter 3

# Experiments

## 3.1 Distillation of saturated aqueous solution of *n*-butanol

### 3.1.1 Purpose

( ⅰ ) Familiar with organic laboratory facilities, learn the installation and disassembly methods of organic chemistry experimental equipment.

( ⅱ ) Familiar with and master the operation of atmospheric distillation, understand the principle and significance of distillation.

### 3.1.2 Principle

The vapor pressure of a liquid increases with increasing temperature. When the vapor pressure of a liquid increase to be equal to the total pressure (usually atmospheric pressure) acting on the liquid surface, many bubbles escape from the liquid, and the liquid begins to boil. The temperature at this time is called the boiling point of the liquid. Obviously, the boiling point is related to external pressure. The unit of vapor pressure is generally expressed in Pa. The boiling point usually refers to the boiling temperature at $1.332\times10^5$ Pa (one atmosphere). For example, the boiling point of water is 100℃, which means that water boils at 100℃ under a $1.332\times10^5$ Pa pressure.

Heating the liquid to boil and turn it into vapor, and then condense the vapor into liquid. The combined operation of these two processes is called distillation. Obviously, distillation can separate volatile and non-volatile substances. Distillation is a method of purifying substances and separating mixtures. The boiling point of compounds can also be measured by distillation so that it can be used for the identification of organic compounds. Distillation can also separate liquid mixtures with different boiling points. In order to obtain a better separation effect, the boiling points of the components of the mixture must differ significantly (at least 30℃). When distillation is performed under atmospheric pressure, because the atmospheric pressure is not precisely $1.332\times10^5$ Pa, adding a correction value to the observed boiling point is strictly necessary. However, the deviation is generally minimal even if the atmospheric pressure differs by 20mmHg, and this correction value does not exceed $\pm 1$ ℃ so that it can be ignored. In the distillation operation, the composition of the original distillation liquid and the distillate composition is different. The composition of the distillate depends on the composition of the gas phase in the distillation process. The measured boiling point is the boiling point of the distillate.

In the distillation process, it is possible to obtain a mixture similar to a simple compound with a fixed boiling point and a fixed composition. In the equilibrium state, the gas phase composition and the liquid phase are the same, and the corresponding temperature at this time is called the azeotropic temperature or azeotropic point. This type of mixed solution can not be separated by normal distillation. The azeotropic temperature is lower than the low boiling point is called the lowest azeotrope; if higher than the high boiling point is called the highest azeotrope.

When the liquid is heated, many bubbles are formed at the bottom of the flask. Air dissolved in the liquid or air adsorbed on the bottle wall contributes to the formation of such bubbles, and the rough surface of the glass also promotes it. Such tiny bubbles (called vaporization centers) can serve as the core of large vapor bubbles. At boiling point, the liquid releases a large amount of vapor into tiny bubbles. When the total pressure in the bubbles increases to exceed the atmospheric pressure and is enough to overcome the pressure generated by the liquid column, the vapor bubbles rise and escape the liquid surface. Therefore, if there are many tiny air bubbles or other vaporization centers in the liquid, the liquid can boil smoothly. If there is almost no air in the fluid, and the bottle wall is very clean and smooth, it is difficult to form bubbles. In this way, the temperature of the liquid may rise much above the boiling point without boiling. This phenomenon is called "overheating". Once a bubble is formed, because the vapor pressure at this temperature exceeds the sum of atmospheric pressure and liquid column pressure, the rising bubble increases very quickly and probably flushes the liquid out of the flask. This abnormal boiling, called "bumping". Therefore, porous zeolites should be added before heating to introduce the gasification center to ensure stable boiling.

Zeolites are generally loose and porous on the surface, which adsorbs air. At any conditions, do not add zeolites to the heated liquid that is close to boiling; otherwise, it will cause danger due to the sudden release of a large amount of vapor, and the liquid ejects from the flask. If you forget to add zeolites before heating, you should remove the heat source, wait until the liquid is cooled to below the boiling point, and add zeolites. If the boiling stops in the middle, new zeolites should be added before reheating. Because the initially added zeolites expel part of the air during heating and adsorb liquid during cooling, it may have lost its effect.

### 3.1.3 Reagents

Sat. aqueous solution of $n$-butanol    50mL

### 3.1.4 Apparatus

The schematic diagram of the distillation experiment device of $n$-butanol saturated aqueous solution is shown in Fig 3-1.

### 3.1.5 Procedure

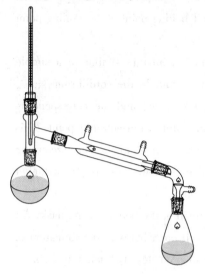

Fig 3-1  The schematic diagram of the distillation of saturated aqueous solution of $n$-butanol device

( i ) Install the distillation instrument according to the above figure. Generally, the iron frame is placed first, starting from the heat source, according to the principle of from bottom to top, from left to right. Use an iron stand to hold the round-bottom flask vertically with iron clamps. Its height should be considered to facilitate the removal of the heating mantle. Do not clamp the flask too tight or too loose, and then the distilling head can be placed directly on the fixed round-bottom flask. Use another iron frame to tilt the straight condenser tube with an iron clamp, adjust its position to be coaxial with the distillation head branch tube, and then connect to it. Connect the thermometer to the screw joint, and adjust the position of the thermometer in the distillation head to keep the same horizontal line of the upper end of the mercury bulb and the lower end of the distillation head branch pipe. The lower end of the tailpipe should be extended into the conical flask used for receiving the distillate, and the conical flask should be open to the atmosphere. The entire device must be installed correctly, whether you view it from the front or side, the axes of each instrument in the set must be in the same plane.

( ii ) In a 100mL round bottom flask, place 50mL of an aqueous solution saturated with $n$-butanol (this solution is dyed blue with methylene blue). When adding, remove the thermometer firstly, and then add the distilled liquid to the round bottom flask through a glass funnel. Be careful not to pour liquid directly from the distillation head.

(iii) Add 2-3 zeolites, then install the thermometer and open the condensate water.

(iv) Use an electric heating mantle to heat. The voltage can be higher initially, and the heating mantle should not be in contact with the round bottom flask directly. Pay attention to the phenomenon in the round bottom flask and the change in the thermometer reading.

(v) When the liquid in the bottle begins to boil, the vapor gradually rises. When the vapor reaches the temperature, the thermometer reading increases sharply. At this time, the voltage should be appropriately lowered to make the temperature drop slightly to keep the droplets on the mercury ball and vapor in balance.

(vi) Then increase the voltage slightly and proceed with distillation. Control the heating and adjust the distillation speed, usually 1-2 drops per second. When the thermometer reading rises to 93℃, use a dry Erlenmeyer flask as the receiver, and collect fractions at 93-95℃.

(vii) When the thermometer reading exceeds 95℃, remove the heating mantle and stop heating.

(viii) Using a separatory funnel, separate the upper oily liquid, measure its milliliters, and calculate the recovery rate.

### 3.1.6 Chart

The experimental process of distillation of saturated aqueous solution of n-butanol is shown in Fig 3-2.

Fig 3-2  The experimental process of distillation of saturated aqueous solution of n-butanol

### 3.1.7 Notes

(i) 100 g of water can dissolve 7.920g of n-butanol at 25℃.

(ii) The flow rate of the condensed water is suitable to ensure that the steam is fully condensed. Usually, only a small flow rate is required.

(iii) Butanol and water form an azeotropic mixture in the gas phase. The boiling point is 93 ℃, and the mass fraction of its composition is 55.5% n-butanol and 44.5% water.

(ⅳ) Distillation of organic solvents should use small mouth receivers, such as conical flasks. The lower end of the connecting pipe should be extended into the conical flask used for receiving the distillate, but the conical flask should be opened to the atmosphere, otherwise it will cause a closed system and cause an explosion.

(Ⅴ) Before using the separatory funnel, check whether the piston leaks and fix it with an iron ring. Dispense after standing. The upper liquid pours out from the upper mouth, and the lower liquid is discharged from the lower mouth.

### 3.1.8 Questions

(ⅰ) What is the boiling point? What is the relationship between the boiling point of a liquid and atmospheric pressure? Is the boiling temperature of a compound recorded in the literature the same as your boiling temperature?

(ⅱ) Why should the amount of liquid in the distillation flask not exceed 2/3 of the volume and not less than 1/3 of the volume during distillation?

(ⅲ) What is the effect of adding zeolite during distillation? If you forget to add zeolite before distillation, can you immediately add zeolite to the nearly boiling liquid? Can the used zeolite continue to be used when it is re-distilled?

(ⅳ) During distillation, if the mercury bulb of the thermometer exceeds the upper edge of the branch pipe of the distillation flask, what effect will it have on the result?

(Ⅴ) When the distillate comes out after heating, it is found that there is no water in the condenser tube. What should you do? Why?

(ⅵ) If a liquid has a constant boiling point, can it be considered a simple substance?

(ⅶ) Can you get pure n-butanol from the upper oily substance obtained from the above experiment of distilling a saturated aqueous solution of n-butanol?

## 3.2 Synthesis of 1-bromobutane

### 3.2.1 Purpose

(ⅰ) Learn the principle and method of preparing 1-bromobutane with sodium bromide, concentrated sulfuric acid, and n-butanol.

(ⅱ) Learn how to use a gas trap apparatus.

## 3.2.2 Principle

$$H_2SO_4 + NaBr \longrightarrow NaHSO_4 + HBr$$
$$CH_3CH_2CH_2CH_2OH + HBr \rightleftharpoons CH_3CH_2CH_2CH_2Br + H_2O$$

---

side reactions:

$$H_2SO_4 + HBr \longrightarrow SO_2 + H_2O + Br_2$$
$$CH_3CH_2CH_2CH_2OH \xrightarrow{H_2SO_4} CH_3CH_2CH_2CH_2OCH_2CH_2CH_2CH_3 + H_2O$$
$$CH_3CH_2CH_2CH_2OH \xrightarrow{H_2SO_4} CH_3CH_2CH_2=CH+CH_3CH=CH_2CH_3 + H_2O$$

The main reaction of this experiment is reversible. In order to improve the yield, usually an excess of HBr is used. On the other hand, NaBr and $H_2SO_4$ are used to make HBr, which can increase the utilization rate of HBr; at the same time, $H_2SO_4$ also plays a role in catalyzing dehydration. During the reaction, in order to prevent the reactant n-butanol and the product 1-bromobutane from escaping from the reaction system, the reaction adopts a reflux device. Since HBr is poisonous and HBr gas is difficult to condense, you should install a gas absorption device to prevent HBr from escaping and polluting the environment. After refluxing, rough distillation is carried out. The product 1-bromobutane is separated, which is convenient for the subsequent separation and purification operation; on the other hand, the rough distillation process can further make the reaction of alcohol and HBr more complete.

The crude product contains unreacted alcohol and ether produced by side reactions, which can be removed by washing with concentrated $H_2SO_4$. Because these two compounds can form salts with concentrated $H_2SO_4$:

$$C_4H_9OH + H_2SO_4 \longrightarrow [C_4H_9\overset{+}{O}H_2]HSO_4^-$$
$$C_4H_9OC_4H_9 + H_2SO_4 \longrightarrow [C_4H_9\underset{H}{\overset{+}{O}}C_4H_9]HSO_4^-$$

If 1-bromobutane contains n-butanol, the head fraction with a lower boiling point is formed during distillation (the boiling point of the azeotropic mixture of 1-bromobutane and n-butanol is 98.6℃, containing 87% of 1-bromobutane, 1-butanol 13%), resulting in a decrease in the yield of refined products.

## 3.2.3 Reagents

1-butanol                          12.4mL (0.136mol)

Anhydrous sodium bromide      16.6g (0.16mol)
Concentrated sulfuric acid    20mL (0.36mol)
10% Aqueous sodium carbonate solution
Anhydrous calcium chloride

### 3.2.4 Apparatus

The schematic diagram of the synthesis of 1-bromobutane device is shown in Fig 3-3.

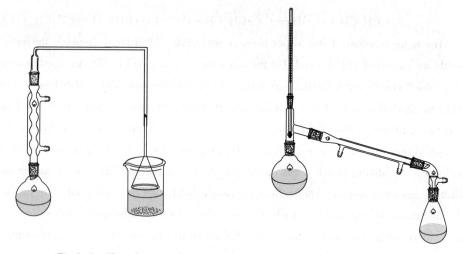

Fig 3-3　The schematic diagram of the synthesis of 1-bromobutane device

### 3.2.5 Procedure

( i ) Place16.6g of powdered sodium bromide and 12.4mL of *n*-butanol in a 100mL round-bottomed flask, add several zeolites, and then fit a reflux condenser to the flask.

( ii ) Dissolve 20mL of concentrated sulfuric acid in 20mL of water, and then allow the acid to dropwise into the flask from the top of the condenser, keeping the contents well shaken meanwhile and occasionally cooled in an ice-water bath.

( iii ) Using an adaptor, connect a gas absorption device to the water condenser when the addition is completed, and then gently boil the mixture for about 40 minutes.

( iv ) Then remove the reflux condense, assemble a distillation apparatus, and distill off the crude *n*-butyl bromide until no more oil drops.

( V ) Purify the distillate by shaking it with 1.5mL of concentrated sulfuric acid in a separator funnel, run off the lower layer of acid, and extract with 1.5mL of concentrated sulfuric acid again. Then run off the lower layer of acid, shake the bromide layer in the funnel cautiously with 10mL of water, 5mL of aqueous sodium carbonate solution (10%), and 10mL water.

(ⅵ) Decant the lower layer of bromide, dry it with anhydrous calcium chloride and finally distill the filtered cure bromide from a 30mL round bottom flask. Collect the n-butyl bromide as a colorless liquid of b.p. 99-102℃.

### 3.2.6 Chart

The experimental process of synthesis of 1-bromobutane is shown in Fig 3-4.

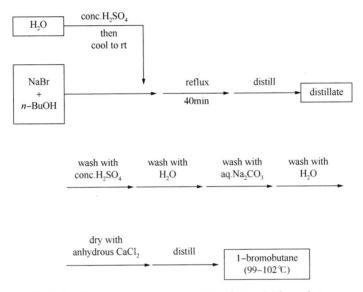

Fig 3-4  The experimental process of synthesis of 1-bromobutane

### 3.2.7 Notes

(ⅰ) When adding concentrated sulfuric acid to water, it is necessary to add a small amount and cool it down while adding.

(ⅱ) The heating of the reaction can not be too drastic; pay attention to the hydrogen bromide absorption device, and do not immerse the glass funnel in water to prevent back suction.

(ⅲ) Pay attention to the order when washing, which layer the product is should be distinguished, and separate the liquid thoroughly.

(ⅳ) When distilling crude products, pay attention to the changes of the oil layer in the reaction mixture and distillate liquid.

( V ) The instrument should be dried during the final distillation, and do not pour the desiccant into the distillation flask.

### 3.2.8  Questions

( i ) Sulfuric acid plays several roles in the reactions, could you describe them?

( ii ) Why is the reflux reaction device used in the 1-bromobutane preparation experiment?

( iii ) Why is the spherical type instead of the straight type condenser tube used as the reflux condenser in the 1-bromobutane preparation experiment?

( iv ) What are the benefits of using 1:1 sulfuric acid in 1-bromobutane preparation experiments?

( V ) When to use a gas absorption device? How to choose absorbent?

( vi ) In the 1-bromobutane preparation experiment, what is the purpose of adding concentrated sulfuric acid to the crude product?

( vii ) Why must the crude n-butyl bromide be dried carefully with calcium chloride before the final distillation?

( viii ) The crude product is distilled in the 1-bromobutane preparation experiment by connecting the condenser tube and the distillation flask with a 75-degree elbow. Can it be changed to a general distillation device for crude distillation? How to control the endpoint of distillation at this time?

( ix ) In this experiment, what will happen if the sulfuric acid concentration is too high or too low?

## 3.3  Synthesis of ethyl acetate

### 3.3.1  Purpose

( i ) Understand the general principles and methods of synthesizing esters from organic acids.

( ii ) Master the operation of distillation and separating funnel.

### 3.3.2  Principle

There are many methods for synthesizing ethyl acetate. For example, it can be prepared by the reaction of acetic acid or its derivatives with ethanol or synthesized by the reaction of sodium acetate with ethyl halide. The most commonly used method is the direct esterification of acetic acid with ethanol under acid catalysis. Concentrated sulfuric acid, hydrogen chloride, p-toluenesulfonic acid, or strong acid cation exchange resins are commonly used as catalysts. If concentrated sulfuric acid is used as the catalyst, the amount is 0.3% of the alcohol. The response is:

$$CH_3COOH + CH_3CH_2OH \xrightarrow{\text{conc. } H_2SO_4} CH_3COOCH_2CH_3$$

---

side-reactions:

$$CH_3CH_2OH \xrightarrow{\text{conc. } H_2SO_4} H_2C=CH_2 + H_2O$$

$$2CH_3CH_2OH \xrightarrow{\text{conc. } H_2SO_4} CH_3CH_2OCH_2CH_3$$

The esterification reaction is reversible, and the measures to increase the yield are: on the one hand, excessive ethanol is added; on the other hand, the product and water are continuously distilled out during the reaction process to promote the equilibrium shift to the direction of ester generation. However, the boiling point of the azeotrope of ester and water or ethanol is close to ethanol. In order to distill out the ester and water produced and try to distill out the ethanol as little as possible, this experiment adopted an excessive amount of ethanol.

### 3.3.3 Reagents

| | |
|---|---|
| Absolute ethanol | 9.5mL (0.08mol) |
| Glacial acetic acid | 6.0g (0.05mol) |
| Concentrated sulfuric acid | 2.5mL (0.075mol) |
| Saturated sodium carbonate solution | |
| Anhydrous magnesium sulfate | |

### 3.3.4 Apparatus

The schematic diagram of the synthesis of ethyl acetate device is shown in Fig 3-5.

### 3.3.5 Procedure

( i ) Add 9.5mL of absolute ethanol and 6mL of glacial acetic acid to a 50mL round bottom flask, carefully add 2.5mL of concentrated sulfuric acid,

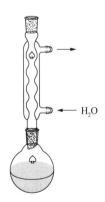

Fig 3-5 The schematic diagram of the synthesis of ethyl acetate device

mix well, add several zeolites, and then install the condenser.

(ii) Carefully heat the reaction mixture and keep it refluxed for 30 minutes. After the reaction mixture in the flask cooled, change the reflux device to a distillation device, and cool the receiving flask with cold water. Heat to distill off ethyl acetate until the volume of the distillate is about 1/2 of the total volume of the reactants.

(iii) Slowly add saturated sodium carbonate solution to the distillate, and keep shaking until no more gas is generated (the pH test paper is not acidic), then transfer the mixed solution to a separatory funnel, and divide the lower aqueous solution.

(iv) Pour the obtained organic layer into a small beaker, dry with an appropriate amount of anhydrous magnesium sulfate, distill the dried solution, and collect fractions at 73-78℃.

### 3.3.6 Chart

The experimental process of synthesis of ethyl acetate is shown in Fig 3-6.

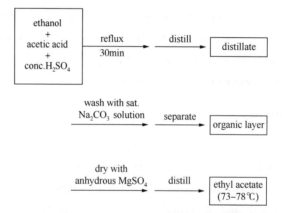

Fig 3-6  The experimental process of synthesis of ethyl acetate

### 3.3.7 Notes

(i) When adding sulfuric acid, add it slowly, shake while adding.

(ii) Pay attention to outgassing during wash with saturated sodium carbonate.

(iii) After washing the organic layer with saturated sodium carbonate, separate the water phase cleanly.

### 3.3.8 Questions

(i) Why use excessive ethanol?

( ii ) What are the main impurities in the crude ethyl acetate? How to remove them?

( iii ) Can sodium hydroxide be used instead of concentrated sodium carbonate for washing? Please explain the reason.

## 3.4 Synthesis of *n*-butyl acetate

### 3.4.1 Purpose

( i ) Learn the principle and method of preparing *n*-butyl acetate by direct esterification of *n*-butyl alcohol with acetic acid.

( ii ) Master the techniques of distillation, extraction, and reflux with a water separator.

### 3.4.2 Principle

The reaction of acid and alcohol to produce ester is reversible:

$$CH_3COOH + n\text{-}C_4H_9OH \xrightarrow{\text{conc. } H_2SO_4} CH_3COO(CH_2)_3CH_3 + H_2O$$

---

side reactions:

$$n\text{-}C_4H_9OH \xrightarrow{\text{conc. } H_2SO_4} (n\text{-}C_4H_9)_2O$$

$$n\text{-}C_4H_9OH \xrightarrow{\text{conc. } H_2SO_4} CH_3CH_2CH_2=CH$$

$$+$$

$$CH_3CH=CH_2CH_3$$

In order to improve the yield, the following measures are adopted generally: ① make a reactant excessive; ② remove a product in the reaction (distill product or water); ③ use special catalysts.

There are three methods for preparing esters directly from acids and alcohols in the laboratory. The first is the azeotropic distillation: the resulting ester and water are distilled as an azeotrope, condensed and separated through a water separator, and the oil layer is returned to the reactor. The second is the extraction esterification method: the solvent is added, so that the solvent and the resulting ester are dissolved in the solvent, and separated from the water layer. The third is the direct reflux method: one kind of reactant is excessive and reflux directly.

It is better to prepare *n*-butyl acetate with azeotropic distillation and water separation method. Esters and acids can form binary or ternary azeotropes, so the azeotrope distillation is chosen to remove the water produced in the reaction. The resulting ester and water are distilled as an azeotrope. After condensation, the water layer is separated through a water trap, and the oil layer is returned to the reactor.

### 3.4.3 Reagents

| | |
|---|---|
| 1-Butanol | 11.5mL (0.125mol) |
| Glacial acetic acid | 7.2mL (0.125mol) |
| Anhydrous sodium bromide | 8.3g (0.08mol) |
| Concentrated sulfuric acid | |
| 10% Aqueous sodium carbonate solution | |
| Anhydrous magnesium sulfate | |

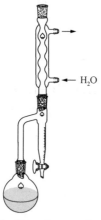

Fig 3-7 The schematic diagram of the synthesis of n-butyl acetate device

### 3.4.4 Apparatus

The schematic diagram of the synthesis of n-butyl acetate device is shown in Fig 3-7.

### 3.4.5 Procedure

(ⅰ) Mix 11.5mL of n-butanol and 7.2mL of glacial acetic acid in a 50mL three-necked bottomer flask. Add cautiously 0.2mL of concentrated sulfuric acid (use a small measuring cylinder or a calibrated dropper pipette). Assemble the apparatus shown in the figure.

(ⅱ) Add a proper amount of water into the water separator with a 2.5-3mL reservation. Install a reflux condenser and reflux the mixture for 60 minutes until the water separator separates no water. Cooling, merge the organic layers.

(ⅲ) Pour the organic mixture into about 10mL of water in a separatory funnel, remove the upper layer of crude ester and rewash it with 10mL of 10% sodium carbonate solution and 10mL of water.

(ⅳ) Dry the crude ester with anhydrous magnesium sulfate, distill the dried solution and collect the fraction at 124-125℃.

### 3.4.6 Chart

The experimental process of synthesis of n-butyl acetate is shown in Fig 3-8.

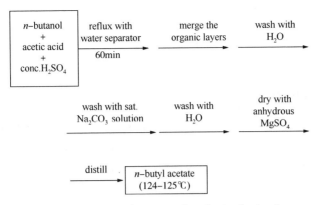

Fig 3-8　The experimental process of synthesis of n-butyl acetate

## 3.4.7　Notes

（ⅰ）When concentrated sulphuric acid is used as a catalyst, a small amount is enough; a solid acid can also be used as a catalyst.

（ⅱ）Vibrate the solution after the sulfuric acid has been added to causes an evenly mixed solution.

（ⅲ）Inspect the cock of the water separator before installing the instrument to be sure it is nimble and sealed.

（ⅳ）The lower component flows through the bottom of the separating funnel when separating the solution while the upper pours from the upper opening.

## 3.4.8　Questions

（ⅰ）How to judge the end of the reaction?

（ⅱ）In the esterification experiment, 2-3mL of water is released in advance when using the water trap. How is this value considered?

（ⅲ）What side reactions are there probably in this experiment? What is the aim of each step of lavation?

（ⅳ）What similarities and differences do extraction and lavation have in principle?

# 3.5　Synthesis of ethyl benzoate

## 3.5.1　Purpose

（ⅰ）Learn the mechanism of the esterification reaction and the preparation of ethyl benzoate.

( ii ) Master the use of water separator and basic operations such as distillation and extraction.

### 3.5.2 Principle

Direct esterification of a carboxylic acid is an important method for preparing carboxylic acid esters in industry and laboratories. The commonly used catalysts include sulfuric acid, hydrogen chloride, and *p*-toluenesulfonic acid.

$$R-\underset{\underset{O}{\|}}{C}-OH + H-OR \xrightleftharpoons{H^+} R-\underset{\underset{O}{\|}}{C}-OR + H_2O$$

The role of the acid is to protonate the carboxyl group to increase the reactivity of the carbonyl group. The reaction is reversible. To promote the reaction in the direction of ester formation, an excess of carboxylic acid or alcohol is usually used, or the ester or water generated in the reaction is removed, or both can be adopted simultaneously. The cyclohexane added in this experiment is used as a water-carrying agent to remove the water generated in the reaction so as to facilitate the forward progress of the reaction. The reaction formula is as follows:

$$C_6H_5COOH + CH_3CH_2OH \xrightarrow{conc.\ H_2SO_4} C_6H_5COOC_2H_5$$

side-reactions:

$$CH_3CH_2OH \xrightarrow{conc.\ H_2SO_4} H_2C=CH_2 + H_2O$$

$$2CH_3CH_2OH \xrightarrow{conc.\ H_2SO_4} CH_3CH_2OCH_2CH_3$$

### 3.5.3 Reagents

| | |
|---|---|
| Benzoic acid | 3.05g (0.025mol) |
| Absolute ethanol | 7.5mL (0.13mol) |
| Concentrated sulfuric acid | 1.0mL |
| Sodium carbonate | |
| Cyclohexane | |
| Diethyl ether | |
| Anhydrous magnesium sulfate | |

### 3.5.4 Apparatus

The schematic diagram of the synthesis of ethyl benzoate apparatus is shown in Fig 3-9.

### 3.5.5 Procedure

( i ) In a 50mL round bottom flask, add 3.05g of benzoic acid, 7.5mL of absolute ethanol, several zeolites, and 1mL of concentrated sulfuric acid. After shaking, install a water separator, reflux condenser, and heat to reflux for 1 hour. Then cool the reaction mixture slightly, add 5-7mL cyclohexane to the mixture (or cyclohexane also can be added into the water separator when installing the device).

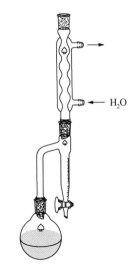

Fig 3-9 The schematic diagram of the synthesis of ethyl benzoate apparatus

( ii ) Heat the flask to reflux again. At the beginning, the reflux speed should be slow. As the reflux progresses, the upper and lower layers of liquid appear in the water separator. When refluxing, allow the upper liquid to return to the reaction flask to prevent the lower liquid from returning to the reaction flask.

( iii ) When the lower layer liquid approaches the branch pipe of the water separator, part of the lower layer liquid is released. Until the upper layer is transparent, no drops of water can be seen. Remove the distillate in the water trap, distill the reaction mixture, let the steam condense to the water trap, separate the condensate, prevent the condensate from flowing back to the flask until distilling most of the ethanol/cyclohexane.

( iv ) The remainder was cooled and poured into a beaker containing 25mL of water. Rinse the reaction flask with a small amount of ethanol, pour this solution into the beaker. Under stirring, add a small amount of sodium carbonate in batches and stir well until no carbon dioxide escapes and the solution is alkaline.

( V ) The mixture was transferred to a separatory funnel, separate the organic layer and extract the aqueous layer with ether (8mL×2). Combine the organic layer and the extract, wash with 10mL saturated sodium chloride solution, dry with anhydrous magnesium sulfate, and leave for at least 15 minutes.

( vi ) Place the dried ether solution into a distillation flask, first evaporate the ether in a water bath, and then heat it on an electric heating mantle to collect the distillate at 210–213℃.

### 3.5.6 Chart

The experimental process of synthesis of ethyl benzoate is shown in Fig 3–10.

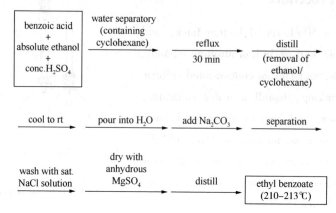

Fig 3–10  The experimental process of synthesis of ethyl benzoate

### 3.5.7 Notes

( ⅰ ) The solvents used in this experiment are flammable. Take care to avoid fire.

( ⅱ ) Pay attention to the safety of taking concentrated sulfuric acid. When adding concentrated sulphuric acid, mix well to prevent carbonization.

( ⅲ ) Because of the toxicity of benzene, cyclohexane is selected as a water-carrying agent instead of benzene in this experiment. Before adding cyclohexane, the heating time should be enough; when adding, the mixture's temperature should be less than 80℃.

( ⅳ ) This experiment can also be carried out according to the following steps: mix 6.0g of benzoic acid, 18mL of absolute ethanol, and 1.2mL of concentrated sulfuric acid. After heating and refluxing for 3 hours, change the reflux device to a distillation device. The treatment method after the evaporation of ethanol is the same as above.

### 3.5.8 Questions

( ⅰ ) What principles and measures are used in this experiment to improve the yield of the equilibrium reaction?

( ii ) In the experiment, how do you use the physical constants of compounds to analyze phenomena and guide operations?

# 3.6 Synthesis of *n*-butyl ether

## 3.6.1 Purpose

( i ) Learn the reaction principle and method of dehydration between alcohol molecules to produce ether.

( ii ) Learn how to use a water-trap vessel.

## 3.6.2 Principle

The dehydration reaction between alcohol molecules is a common method for preparing simple ethers. Under the catalysis of acid, the hydroxyl group of the alcohol is protonated, which increases the electrophilicity of α-C and the leaving property of the hydroxyl group, making it more prone to bimolecular nucleophilic substitution and loss of protons to obtain ethers. The reaction is generally carried out at 135℃. If the temperature is too high (more than 150℃), an elimination reaction will occur, and olefins will be obtained.

The formation of ether is reversible, and you should distill water or ether continuously during the reaction to make the reaction proceed toward the formation of ether. In this experiment, the continuous steaming of water is achieved by using a water trap.

$$CH_3CH_2CH_2CH_2OH \xrightleftharpoons[134-135℃]{H_2SO_4} CH_3CH_2CH_2CH_2OCH_2CH_2CH_2CH_3 + H_2O$$

---

side reactions:

$$CH_3CH_2CH_2CH_2OH \xrightarrow[>135℃]{H_2SO_4} CH_3CH_2CH_2=CH + CH_3CH=CH_2CH_3 + H_2O$$

## 3.6.3 Reagents

| | |
|---|---|
| 1-butanol | 15.5mL (0.17mol) |
| Concentrated sulfuric acid | 2.5mL (0.0475mol) |
| 5% sodium hydroxide solution | |
| Saturated calcium chloride solution | |
| Anhydrous calcium chloride | |

### 3.6.4 Apparatus

The schematic diagram of the synthesis of n-butyl ether apparatus is shown in Fig 3-11.

### 3.6.5 Procedure

( i ) Place 15.5mL of n-butanol and 2.5mL of concentrated sulfuric acid in a 50mL round-bottomed flask, add several zeolites, fit a water-trap vessel to the flask, and then connect the water condenser to the water-trap vessel.

( ii ) Heat it to a slight boiling point and reflux. After about 60 minutes, the temperature of the reaction solution can reach 134-136℃. If you continue to heat, the reaction solution will turn black, and a by-product of butene will be obtained.

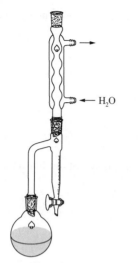

Fig 3-11 The schematic diagram of the synthesis of n-butyl ether apparatus

( iii ) Pour the reaction solution into a separatory funnel containing 25mL of water. Separate the lower layer of water, wash the upper crude product with 10mL of water, 10mL of 5% sodium hydroxide solution, 10mL of water, 10mL of saturated calcium chloride solution, and dry with anhydrous calcium chloride.

( iv ) Finally, distill the crude ether in a 50mL round bottom flask, and collect the n-butyl ether as a colorless liquid of b.p. 139-142℃.

### 3.6.6 Chart

The experimental process of synthesis of n-butyl ether is shown in Fig 3-12.

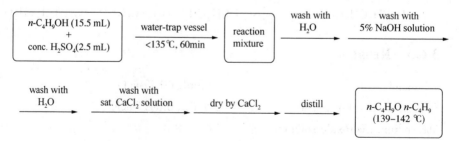

Fig 3-12 The experimental process of synthesis of n-butyl ether

### 3.6.7 Notes

( i ) When adding, if the *n*-butanol and concentrated sulfuric acid are not well shaken and mixed thoroughly, the sulfuric acid is too concentrated. The reaction solution will turn black after heating.

( ii ) When the reaction started to reflux, the temperature could not reach 135℃ immediately because of the presence of azeotropes. However, when the water is distilled out, the temperature rises gradually and finally reaches above 135℃, and at this time, you should stop the heating. If the temperature rises too high, the reaction solution will be carbonized, and a large amount of by-product butene will be generated.

( iii ) *n*-Butanol is soluble in saturated calcium chloride solution, while *n*-butyl ether is slightly soluble.

### 3.6.8 Questions

( i ) How to strictly control the reaction temperature during the reaction?

( ii ) How to judge that the reaction is relatively complete?

## 3.7 Synthesis of acetanilide

### 3.7.1 Purpose

( i ) Learn the principle and method of preparing acetanilide by acetic acid as the acetylation reagent.

( ii ) Master the technique of fractional distillation and recrystallizing solids.

### 3.7.2 Principle

$$C_6H_5-NH_2 + CH_3COOH \longrightarrow C_6H_5-NHCOCH_3 + H_2O$$

### 3.7.3 Reagents

| | |
|---|---|
| Acetic acid | 8.5mL (0.13mol) |
| Aniline | 5mL (0.055mol) |
| Zinc powder | |
| Charcoal | |

### 3.7.4 Apparatus

The schematic diagram of the synthesis of acetanilide apparatus is shown in Fig 3-13.

### 3.7.5 Procedure

( i ) Pour 5mL of aniline into a 50mL conical flask, and 8.5mL of acetic acid, and about 0.1g zinc powder to the flask.

( ii ) Assemble a fractional column with a thermometer and a condenser, and use a 50mL flask as a collector.

( iii ) Heat the mixture to reflux for 45 mins, and the temperature of the distilling vapor should not exceed 105℃.

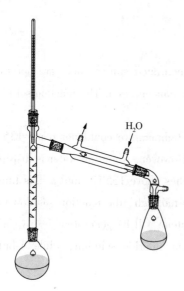

Fig 3-13  The schematic diagram of the synthesis of acetanilide apparatus

( iv ) Then pour the hot liquid into 100mL of cold water in a beaker with stirring. In a few seconds, the precipitation of acetanilide forms.

( v ) Collect the acetanilide by filtration, wash it well with water, and to recrystallize from water. M.p. 113℃.

### 3.7.6 Chart

The experimental process of synthesis of acetanilide is shown in Fig 3-14.

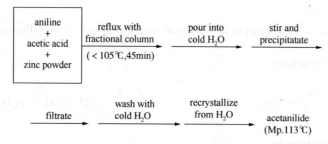

Fig 3-14  The experimental process of synthesis of acetanilide

### 3.7.7 Notes

( i ) Freshly distilled aniline should be used to get good results, or a small amount of zinc can be added to the reaction mixture. Zinc reduces the colored impurities in the aniline and also prevents its oxidation during the reaction.

(ii) Prolonged heating and the use of an excess of acetic anhydride should be avoided.

(iii) The reaction mixture should be cooled and then poured in ice-cold water; otherwise, hydrolysis of acetanilide may occur.

(iv) Aniline is toxic and can be absorber through the skin. Use in a fume hood.

(V) Handle acetic anhydride carefully as they irritate the eyes.

### 3.7.8 Questions

(i) Why must the temperature of the distilling vapor be controlled to about 105°C?

(ii) How to judge the termination of this reaction?

(iii) How many milliliters of water are produced during the preparation reaction of acetanilide, and why is the collected liquid far more than the theoretical amount?

(iv) What are the characteristics of acylation with acetic acid and acylation with acetic anhydride? Besides, what are the acylating reagents?

(V) What principle is used in this experiment to increase the yield of acetanilide?

## 3.8 Recrystallization of acetanilide

### 3.8.1 Purpose

(i) Learn the principles and methods of recrystallization.

(ii) Learn the choice of solvent and the determination of the amount, and understand the commonly used solvents and mixed solvents for recrystallization.

### 3.8.2 Principle

Most compounds have different solubility at different temperatures. A crude compound is dissolved in a suitable solvent at a higher temperature, and then filtered to remove insoluble impurities and colored impurities (decolorization can be performed with activated carbon). The filtrate is then cooled in ice water, the product crystallizes, and then filter to make the pure compound, while the soluble impurities remain in the solution, allowing the solid material to be purified.

The solubility of acetanilide in water at different temperatures is list in Table 3-1.

Table 3-1  The solubility of acetanilide in water at different temperatures

| T/°C | 25 | 80 | 100 |
|---|---|---|---|
| Solubility/(g/100mL $H_2O$) | 0.563 | 3.5 | 5.2 |

### 3.8.3  Reagents

Acetanilide　　　　　　　　　　3.4g (0.136mol)
Activated carbon
Ethanol

### 3.8.4  Apparatus

The schematic diagram of recrystallization of acetanilide apparatus is shown in Fig 3-15.

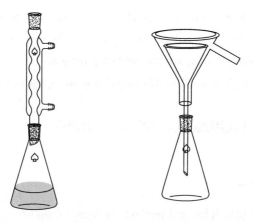

Fig 3-15  The schematic diagram of recrystallization of acetanilide apparatus

### 3.8.5  Procedure

( ⅰ ) In a 250mL conical flask, add the crude acetanilide and an appropriate amount of water, and then install a reflux device.

( ⅱ ) Heat to boiling, add a small amount of water from the upper mouth of the condenser, and heat to boiling again to dissolve the acetanilide (If there are undissolved oil droplets, add an appropriate amount of water to dissolve it until there are no oil droplets completely).

( ⅲ ) After cooling slightly, add 15% of the total amount of water, add a small amount of powdered activated carbon, and continue heating to make it boil for 1-2min.

( ⅳ ) Use the pre-prepared insulation funnel and folded filter paper to filter while it is hot.

( ⅴ ) Let stand to cool the filtrate, filter with suction after the crystals are separated, and squeeze the filter cake with a glass plug to remove the water in the crystals as much as possible. Place the product in a watch glass to dry.

### 3.8.6  Chart

The experimental process of recrystallization of acetanilide is shown in Fig 3-16.

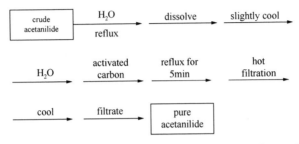

Fig 3-16  The experimental process of recrystallization of acetanilide

### 3.8.7  Notes

( ⅰ ) Pay attention to the amount of recrystallization solvent; when adding solvent, pay attention to avoid fire.

( ⅱ ) When adding activated carbon, the solution should be cooled slightly before adding to avoid bumping. After adding activated carbon, it needs to be heated slowly to prevent excessive boiling. In addition, do not add a large amount of activated carbon at one time to avoid loss of adsorption products.

( ⅲ ) During hot filtration, be careful not to make too much liquid in the funnel to avoid overflow or clogging of the funnel by precipitation of crystals.

### 3.8.8  Questions

( ⅰ ) How to calculate the amount of solvent?

( ⅱ ) What methods are generally used to purify solid organic compounds? How to use a simple way to identify the purity of solid organic compounds?

( ⅲ ) Which steps are generally included in the recrystallization method? What is the main purpose of each step?

( ⅳ ) What are the properties of the most suitable solvent for recrystallization of an organic compound?

( V ) When heating the crude product to be recrystallized from the solvent, why add a little less volume meter of the solvent? Then add gradually until it just dissolves?

( vi ) Why can't activated carbon be added when the solution is boiling?

( vii ) When using organic solvents to recrystallize, which operations are easy to catch fire? How can it be avoided?

## 3.9 Synthesis of *p*-bromoacetanilide

### 3.9.1 Purpose

( i ) Learn the methods and principles of aromatic halogenation reactions.

( ii ) Master the aromatics oxidation bromination in green chemistry, and compare the advantages and disadvantages of traditional methods.

( iii ) Consolidate the operation technology of recrystallization and melting point determination.

### 3.9.2 Principle

$$NaBr + H_2SO_4 \longrightarrow HBr + NaHSO_4$$
$$2HBr + H_2O_2 \longrightarrow Br_2 + H_2O$$

$$\text{C}_6\text{H}_5\text{NHCOCH}_3 + Br_2 \longrightarrow p\text{-Br-C}_6\text{H}_4\text{NHCOCH}_3 + HBr$$

------------------------------------------------------------

*Total*:

$$\text{C}_6\text{H}_5\text{NHCOCH}_3 + NaBr + H_2SO_4 + H_2O_2 \longrightarrow p\text{-Br-C}_6\text{H}_4\text{NHCOCH}_3 + H_2O + NaHSO_4$$

### 3.9.3 Reagents

Acetanilide                      3.4g (0.025mol)

NaBr                             2.6g (0.0252mmol)

| | |
|---|---|
| 33% $H_2O_2$ | 6.0mL |
| 95% Ethanol | 15mL |
| Conc. $H_2SO_4$ | 2mL |

### 3.9.4 Apparatus

The schematic diagram of synthesis of *p*-bromoacetanilide apparatus is shown in Fig 3-17.

### 3.9.5 Procedure

( ⅰ ) Fit a 100mL three-necked flask with a constant pressure dropping funnel, a reflux condenser, and an electric stirrer.

( ⅱ ) Add 3.4g acetanilide, 15mL of 95% ethanol, 6mL of 33% $H_2O_2$, and 2.6g of NaBr into the three-necked flask. At room temperature, 2mL of $H_2SO_4$ is added dropwise to the flask. The rate of adding is appropriate for the color of bromine formation to fade quickly.

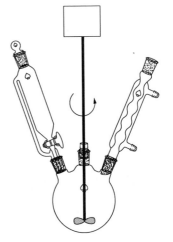

Fig 3-17  The schematic diagram of synthesis of *p*-bromoacetanilide apparatus

( ⅲ ) After the addition, continue to stir for 5 to 10 minutes. Stop stirring and let it cool naturally to precipitate crystals.

( ⅳ ) After cooling, filter with suction, wash the filter cake with cold water, drain it, and place it in the air to dry naturally to obtain larger white needle-like crystals.

( ⅴ ) At the same time, many crystals appear in the filtrate and then filter with suction, wash the filter cake with cold water. After being placed in the air to dry naturally, needle-like crystals with slightly colored fine particles are obtained.

( ⅵ ) Weigh the crystal products, and calculate the yield of the reaction based on the total mass.

### 3.9.6 Chart

The experimental process of synthesis of *p*-bromoacetanilide is shown in Fig 3-18.

### 3.9.7 Notes

( ⅰ ) Concentrated sulfuric acid is highly corrosive, and you must be careful when handling it.

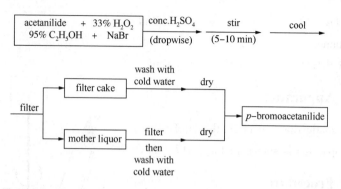

Fig 3-18 The experimental process of synthesis of p-bromoacetanilide

(ii) During the reaction process, add concentrated sulfuric acid slowly.

### 3.9.8 Questions

(i) Why do you need to control the addition rate of concentrated sulfuric acid during the reaction?

(ii) What impurities may exist in the product, and how to remove them?

(iii) In this bromination reaction, what effect does the reaction temperature have on the reaction result?

## 3.10 Synthesis of benzyl alcohol and benzoic acid

### 3.10.1 Purpose

(i) Understand the basic principles of the Cannizzaro reaction.

(ii) Proficiency in basic operations such as extraction, washing, distillation of low-boiling and high-boiling substances.

### 3.10.2 Principle

$$2\ \text{C}_6\text{H}_5\text{-CHO} \longrightarrow \text{C}_6\text{H}_5\text{-COONa} + \text{C}_6\text{H}_5\text{-CH}_2\text{OH}$$

$$\xrightarrow{\text{conc. HCl}} \text{C}_6\text{H}_5\text{-COOH}$$

### 3.10.3 Reagents

| | |
|---|---|
| Benzaldehyde | 21mL (0.2mol) |
| Potassium hydroxide | 18g (0.32mol) |

Diethyl ether

10% sodium carbonate solution

Anhydrous magnesium sulfate

Concentrated hydrochloric acid

### 3.10.4  Apparatus

The schematic diagram of synthesis of benzyl alcohol and benzoic acid apparatus is shown in Fig 3-19.

### 3.10.5  Procedure

( i ) Dissolve 18g of potassium hydroxide in 18mL of water in a beaker or conical flask. Pour the solution into a 250mL reagent bottle, and add 21g of pure benzaldehyde; shake the mixture until it is converted into a thick emulsion. Allow the mixture to stand overnight for 24 hours.

Fig 3-19  The schematic diagram of synthesis of benzyl alcohol and benzoic acid apparatus

( ii ) Add about 60mL of water to the bottle to dissolve the potassium benzoate. Pour the liquid into a separatory funnel, rinse out the bottle with 20mL of ether and add this ether to the solution in the funnel. Shake the solution to extract the benzyl alcohol with the ether, separate the upper ether layer, and extract the lower aqueous solution with the ether two times (each with about 20mL ether).

( iii ) Combine the ether extracts and shake the ether solution twice with 5mL of saturated sodium metabisulphite solution. Then wash it with 10mL of 10% sodium carbonate solution, 10mL of water, dry with anhydrous magnesium sulfate or anhydrous potassium carbonate. After distilling the ether, replace the water condenser with an air condenser. Collect the benzyl alcohol at 204-207℃.

( iv ) Pour the aqueous solution remaining from the ether extraction, stir with a mixture of 50mL concentrated hydrochloric acid, 50mL of water, and about 75g of crushed ice. Filter the precipitated benzoic acid, wash it with cold water, and recrystallize from boiling water.

### 3.10.6  Chart

The experimental process of synthesis of benzyl alcohol and benzoic acid is shown in Fig 3-20.

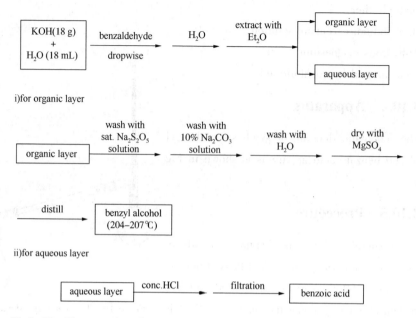

Fig 3-20  The experimental process of synthesis of benzyl alcohol and benzoic acid

## 3.10.7  Notes

( i ) If the reaction cannot be stirred sufficiently in the first step, it will affect the yield of the subsequent reaction. If mix well, the smell of benzaldehyde smell disappears.

( ii ) When separating with a separatory funnel, separate the aqueous layer from below, and decant the organic layer from the top of the funnel; otherwise, it will affect the subsequent operations.

( iii ) Note that the ether extract in the round-bottom flask cannot be heated in a water bath. That is evaporation, not distillation, and it is an error operation.

( iv ) If the water layer is not acidified completely, the benzoic acid will not fully precipitate, leading to the disappearance of the product.

## 3.10.8  Questions

( i ) The Cannizzaro reaction occurs much more slowly in dilute sodium hydroxide solution than in concentrated solution. Why?

( ii ) By what means would aqueous sodium bisulfite remove unchanged benzaldehyde from the reaction mixture?

## 3.11 Synthesis of α-furyl methanol and α-furoic acid

### 3.11.1 Purpose

( i ) Learn the principle and method of preparing α-furoic acid and α-furyl methanol from furfural.

( ii ) Deepen the understanding of the Cannizzaro reaction.

### 3.11.2 Principle

$$2 \text{ furyl-CHO} \longrightarrow \text{furyl-COONa} + \text{furyl-CH}_2\text{OH}$$
$$\xrightarrow{\text{conc.HCl}} \text{furyl-COOH}$$

### 3.11.3 Reagents

| | |
|---|---|
| Furfural | 3.28mL (0.04mol) |
| Sodium hydroxide | 1.6g (0.32mol) |
| Diethyl ether | |
| Anhydrous magnesium sulfate | |
| Concentrated hydrochloric acid | |

### 3.11.4 Apparatus

The schematic diagram of synthesis of α-furyl methanol and α-furoic acid apparatus is shown in Fig 3-21.

### 3.11.5 Procedure

( i ) Dissolve 1.6g of sodium hydroxide in 2.4mL of water in a 50mL beaker, keep the temperature at 8–12℃, and add 3.8g of freshly distilled furfural. After adding, allow the reaction to react for 40 min.

Fig 3-21 The schematic diagram of synthesis of α-furyl methanol and α-furoic acid apparatus

(ii) Add water (about 5mL) to dissolve the sodium 2-furoate. Pour the liquid into a separatory funnel, rinse out the bottle with 3mL of ether and add this ether to the solution in the funnel. Shake the solution to extract the α-furyl methanol with the ether, separate the upper ether layer, and extract the lower aqueous solution with the ether two times (each with about 3mL ether).

(iii) Combine the ether extracts and shake the ether solution twice with 5mL of saturated sodium metabisulphite solution. Wash it with 5mL of 10% sodium carbonate solution, 5mL of water, and dry with anhydrous magnesium sulfate or anhydrous potassium carbonate. After distilling the ether, replace the water condenser with an air condenser. Collect the α-furyl methanol at 168–172℃.

(iv) Pour the aqueous solution remaining from the ether extraction, stir with a mixture of 5mL concentrated hydrochloric acid, 5mL of water, and about 10g of crushed ice. Filter the precipitated 2-furoic acid, wash it with cold water, recrystallize from boiling water.

### 3.11.6  Chart

The experimental process of synthesis of α-furyl methanol and α-furoic acid is shown in Fig 3-22.

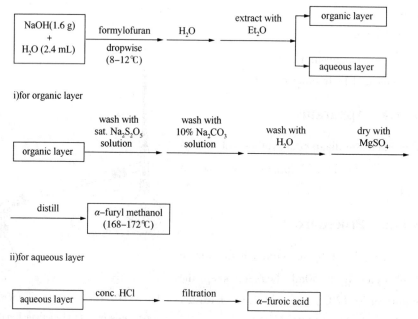

Fig 3-22  The experimental process of synthesis of α-furyl methanol and α-furoic acid

### 3.11.7 Notes

( i ) Furfural is readily oxidized by air and should be distilled again before use.

( ii ) If the reaction temperature is lower than 8℃, the reaction is too slow; if it is higher than 12℃, the reaction temperature will rise quickly, which is difficult to control, and the reactants will turn dark red.

( iii ) When distilling, use a low-boiling liquid distillation device to distill the ether. When the temperature rises to 140℃, use an air condenser to distill again.

### 3.11.8 Questions

( i ) What is the key point of this experiment?

( ii ) Why is it necessary to stir well during the reaction?

## 3.12　Synthesis of benzyl alcohol

### 3.12.1　Purpose

( i ) Understand the principle of phase transfer catalytic reaction and prepare benzyl alcohol by phase transfer catalytic reaction.

( ii ) Practice the use of high-temperature distillation and air condenser.

### 3.12.2　Principle

The preparation of benzyl alcohol with benzyl chloride is a practical example of the production of alcohols by the hydrolysis of halogenated hydrocarbons. The hydrolysis is carried out in an alkaline aqueous solution. Because halogenated hydrocarbons are insoluble in water, this two-phase reaction proceeds very slowly and requires vigorous stirring. If a phase transfer catalyst (PTC) such as tetraethylammonium bromide is added, the reaction time can be significantly shortened. The reaction formula is:

$$2\ \text{C}_6\text{H}_5\text{—CH}_2\text{Cl} + \text{K}_2\text{CO}_3 + \text{H}_2\text{O} \xrightarrow[\text{reflux}]{\text{PTC}} 2\ \text{C}_6\text{H}_5\text{—CHO} + 2\text{KCl} + \text{CO}_2\uparrow$$

### 3.12.3　Reagents

| | |
|---|---|
| Benzyl chloride | 9.5mL (10.1g, 0.08mol) |
| Potassium carbonate | 8.0g (0.06mol) |
| Tetraether ammonium bromide | 2mL (50% solution) |

Diethyl ether
Anhydrous magnesium sulfate

### 3.12.4 Apparatus

The schematic diagram of synthesis of benzyl alcohol apparatus is shown in Fig 3-23.

### 3.12.5 Procedure

( i ) Fit a 250mL three-necked flask with a thermometer, a reflux condenser, and a stirrer.

( ii ) Place 8.0g potassium carbonate, 80mL $H_2O$, and 2mL phase transfer catalyst (PTC), and 9.5mL benzyl chloride into the flask.

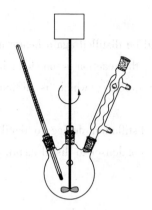

Fig 3-23　The schematic diagram of synthesis of benzyl alcohol apparatus

( iii ) Reflux the mixed liquid gently for about 90min.

( iv ) Stop heating and transfer the reaction mixture to a separator funnel until the mixture is cooling to 30-40℃.

( v ) Separate the organic layer (upper layer), extract the aqueous layer four times by 5mL diethyl ether, and combine the organic and diethyl ether layers. Dry them through anhydrous magnesium sulfate (Do not use absolute calcium chloride).

( vi ) Distill the dried mixed solution under atmospheric pressure. After the diethyl ether is distilled, continue to distill benzyl alcohol under atmospheric pressure until the temperature reaches 200-208℃.

### 3.12.6 Chart

The experimental process of synthesis of benzyl alcohol is shown in Fig 3-24.

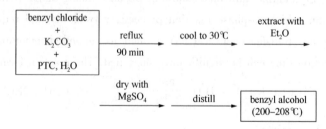

Fig 3-24　The experimental process of synthesis of benzyl alcohol

### 3.12.7 Notes

( i ) Benzyl chloride is tangy, do not touch your skin.

( ii ) The phase transfer catalyst of tetraethylammonium bromide can be replaced by triethyl benzyl ammonium chloride.

( iii ) Although a phase transfer catalyst is used, stirring is still required to accelerate the speed of phase transfer during the reaction.

( iv ) Diethyl ether is flammable and explosive.

( v ) Alcohol cannot be dried by absolute calcium chloride.

( vi ) When distilling the benzyl alcohol, you must use an air condensator. During high-temperature distillation, the temperature of the air condenser is high, so be careful to prevent burns.

( vii ) After the reaction is over, it is advisable to cool to 30–40℃. If the temperature is too low, solids will precipitate, which will affect the next separation operation.

### 3.12.8 Questions

( i ) Why do you use tetraethylammonium bromide as a phase transfer catalyst?

( ii ) Can you synthesis benzyl alcohol with other methods?

( iii ) What is the reaction process of benzyl chloride alkaline hydrolysis?

( iv ) Briefly describe the principle of phase transfer catalytic reaction.

( v ) What is the role of ether extraction? Can other solvents be used instead?

## 3.13 Synthesis of benzoic acid

### 3.13.1 Purpose

( i ) Learn about the oxidation reaction on the benzene ring branch.

( ii ) Master the methods of vacuum filtration and recrystallization purification.

### 3.13.2 Principle

The oxidation reaction is a common method for preparing carboxylic acids. Aromatic carboxylic acids are usually prepared by the oxidation of aromatic hydrocarbons containing α-H. The benzene ring of aromatic hydrocarbons is relatively stable and difficult to oxidize, but the branched chain on the ring will be oxidized to a carboxyl group regardless of its length.

The preparation of carboxylic acid adopts relatively strong oxidation conditions. The oxidation reaction is generally exothermic, so it is crucial to control the temperature of this reaction. If the reaction is out of control, the product will be destroyed, and the yield will decrease, and sometimes there will be a danger of explosion.

$$\underset{}{\text{C}_6\text{H}_5\text{CH}_3} \xrightarrow[\text{H}_2\text{O, reflux}]{\text{KMnO}_4} \underset{}{\text{C}_6\text{H}_5\text{COOK}} \xrightarrow{\text{conc.H}_2\text{SO}_4} \underset{}{\text{C}_6\text{H}_5\text{COOH}}$$

### 3.13.3 Reagents

| | |
|---|---|
| Toluene | 2.7mL (2.3g, 0.025mol) |
| Potassium permanganate | 8g (0.05mol) |
| Diglyme (4wt% in water) | 4mL (0.00125mol) |
| Saturated NaHSO$_3$ solution | |

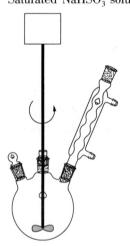

Fig 3–25 The schematic diagram of synthesis of benzoic acid apparatus

### 3.13.4 Apparatus

The schematic diagram of synthesis of benzoic acid apparatus is shown in Fig 3–25.

### 3.13.5 Procedure

( i ) Place 8.0g KMnO$_4$ and 100mL water into a three-neck 250mL round-bottom flask equipped with a mechanical stirrer, a reflux condenser, and a thermometer. Turn on the motor at a low speed to dissolve KMnO$_4$. Add the aqueous solution of toluene with diglyme into the flask. Heat to boiling and reflux for 40min with vigorous stirring. When the reaction is finished, turn off the heat and allow the mixture to cool down.

( ii ) With vigorous stirring, carefully add 10mL of saturated NaHSO$_3$ aqueous solution into the flask from the top of the condenser. If the upper layer of the mixture is still pink, add an aqueous solution of NaHSO$_3$ until the upper layer becomes colorless. Set up an apparatus for vacuum filtration, filter off the precipitate of MnO$_2$.

( iii ) Transfer the filtrate to a clean beaker and acidify the solution by adding concentrated HCl. The benzoic acid product forms as a white precipitate. Collect the product by vacuum filtration, press the crystalline solid as dry as possible. Weigh the product and calculate the percentage yield.

## 3.13.6 Chart

The experimental process of synthesis of benzyl acid is shown in Fig 3-26.

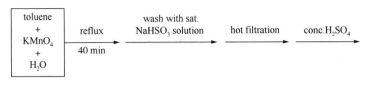

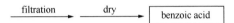

Fig 3-26  The experimental process of synthesis of benzyl acid

## 3.13.7 Notes

( i ) KMnO$_4$ should be added in small portions to prevent the solution from ejecting from the flask.

( ii ) If the filtrate is purple, add a small amount of sodium bisulfite to the filtrate to make the purple color fade before filtration.

( iii ) The solubility of benzoic acid in water increases with increasing temperature.

## 3.13.8 Questions

( i ) In this oxidation reaction, what are the main factors affecting the production of benzoic acid?

( ii ) After the reaction, what is the purpose of adding sodium hydrogen sulfite to the reaction mixture?

( iii ) What are the methods for preparing benzoic acid?

# 3.14  Synthesis of cyclohexene

## 3.14.1 Purpose

( i ) Learn the principle and method of preparing cyclohexene by dehydration of cyclohexanol.

( ii ) Master the basic operating skills of fractional distillation, water bath heating, separation, drying, etc.

### 3.14.2 Principle

$$\underset{\text{OH}}{\bigcirc} \xrightarrow{H_3PO_4} \bigcirc$$

### 3.14.3 Reagents

| | |
|---|---|
| Cyclohexanol | 10.4mL (0.1mol) |
| Concentrated phosphoric acid | 4mL |
| Sodium chloride | |
| 5% sodium carbonate solution | |
| Anhydrous calcium chloride | |

### 3.14.4 Apparatus

The schematic diagram of synthesis of cyclohexene apparatus is shown in Fig 3-27.

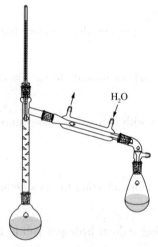

Fig 3-27 The schematic diagram of synthesis of cyclohexene apparatus

### 3.14.5 Procedure

( i ) Place 10.4mL of cyclohexanol, 4mL of concentrated phosphoric acid, and several zeolites into a 50mL dry round-bottom flask.

( ii ) Assemble a fractional column with a thermometer, and use a 50mL Erlenmeyer flask as a collector, and immerse it into ice water.

( iii ) Heat the solution slowly to boil for 60 mins, and the temperature of the distilling vapor should not exceed 73℃.

( iv ) Add 1g of NaCl to the collected distillate in the receiver until the water layer is saturated, and then 5% sodium carbonate solution to neutralize the small amount of acid.

( v ) Pour the liquid into a separatory funnel, and allow the layers to separate. Decant the organic layer from the top of the funnel into a dry Erlenmeyer flask. Add 1-2g of anhydrous calcium chloride to the Erlenmeyer flask, and shake it occasionally to dry the product thoroughly. A clear liquid of cyclohexene is obtained.

(ⅵ) Dry all glassware completely, assemble the distillation apparatus again. Decant the crude product into a round-bottom flask. Add 2-3 zeolites to the flask, and use a 25mL Erlenmeyer flask as a collector that immerses into ice water. Distill and collect the fraction of 82-85℃.

### 3.14.6 Chart

The experimental process of synthesis of cyclohexene is shown in Fig 3-28.

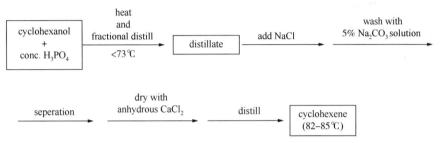

Fig 3-28  The experimental process of synthesis of cyclohexene

### 3.14.7 Notes

( ⅰ ) Cyclohexene is highly flammable; ensure that all joints in the apparatus are tightly mated. If possible, flameless heating is preferred.

( ⅱ ) With the slowly heating, the temperature will not exceed 73℃.

( ⅲ ) The solubility of cyclohexene in the aqueous layer can be decreased in saturated sodium chloride solution.

( ⅳ ) The temperature at the thermometer begins to drop at this experiment. It indicates the end of the reaction.

### 3.14.8 Questions

( ⅰ ) What is the advantage of $H_3PO_4$ as a catalyst in comparison to $H_2SO_4$?

( ⅱ ) In this experiment, why was anhydrous calcium chloride used as the desiccant in this?

## 3.15  Synthesis of cyclohexanone

### 3.15.1 Purpose

( ⅰ ) Master skillful techniques of mechanical stirring, extraction, and distillation.

( ii ) Understand the function of salting-out agents used in layer separation.

### 3.15.2 Principle

$$\text{C}_6\text{H}_{11}\text{OH} + \text{NaOCl} \xrightarrow[\text{CH}_3\text{COOH}]{\text{H}_2\text{O}} \text{C}_6\text{H}_{10}\text{O}$$

### 3.15.3 Reagents

| | |
|---|---|
| Cyclohexanol | 10.4mL (0.1mol) |
| Aqueous sodium hypochlorite solution | 80mL |
| Glacial acetic acid | 25mL |
| Methyl $t$-butyl ether | 25mL |
| Saturated sodium bisulfite solution | |
| KI-starch test paper | |
| Sodium bicarbonate | |
| Anhydrous sodium carbonate | |
| Anhydrous magnesium sulfate | |

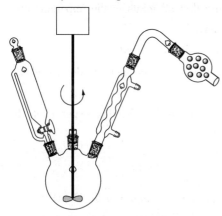

Fig 3-29 The schematic diagram of synthesis of cyclohexanone apparatus

### 3.15.4 Apparatus

The schematic diagram of synthesis of cyclohexanone apparatus is shown in Fig 3-29.

### 3.15.5 Procedure

( i ) Place 10.4mL of cyclohexanol and 25mL of glacial acetic acid into a three-neck 250mL round-bottom flask equipped with a mechanical stirrer, a dropping funnel, a reflux condenser, and a thermometer. A drying tube with granular sodium bicarbonate connects to the top end of the condenser. Decant the sodium hypochlorite solution into the dropping funnel.

( ii ) With stirring, drop the sodium hypochlorite solution into the three-neck flask at a speed that maintains the temperature between 30-35℃. After the addition of sodium hypochlorite solution, the color of the reaction mixture becomes yellow-green color.

Continues stirring for 5min, and test the aqueous layer with KI-starch test paper. If the KI-starch test paper does not turn deep blue, add the sodium hypochlorite solution (ca. 5mL) until a positive test for excess sodium hypochlorite solution is observed.

(iii) Stir the resulting solution for another 15min. Then, add the saturated sodium bisulfite solution to the reaction mixture until the aqueous layer of the solution gives a negative KI-starch test.

(iv) Add 60mL of water and several zeolites to the flask, assemble a simple distillation apparatus, distill the mixture and collect the distillate below 100°C.

(v) Add anhydrous sodium carbonate in small portions to the distillate with stirring until carbon dioxide ceases completely.

(vi) Pour the solution into a funnel, add sodium chloride to the funnel and let the water layer be saturated with sodium chloride. Decant the organic layer from the top of the funnel, wash the aqueous layer with 25mL methyl *tert*-butyl ether, and combine the organic layers. Dry the organic layer with anhydrous magnesium sulfate.

(vii) Dry all glassware completely, decant the dried liquid into a dry 50mL round-bottom flask. Add 2-3 zeolites to this flask, distill and collect the fraction of 150-155°C.

## 3.15.6 Chart

The experimental process of synthesis of cyclohexanone is shown in Fig 3-30.

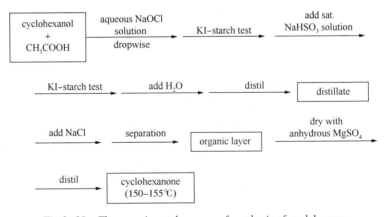

Fig 3-30  The experimental process of synthesis of cyclohexene

## 3.15.7 Notes

(i) Do not allow hypochlorite to contact your skin or eye.

(ii) Sodium bicarbonate can absorb the chlorine gas possibly released from the reaction.

(iii) Cyclohexanone can co-distill with water at 95℃. The fraction below 100℃ mainly contains cyclohexane, water, and a small amount of acetic acid.

(iv) Saturating the aqueous layer with sodium chloride will decrease the solubility of cyclohexanone in water, therefore improving the separation of the organic and water layers.

### 3.15.8 Questions

(i) In addition to sodium bicarbonate solid, are there any methods to absorb chlorine?

(ii) Why must the saturated sodium bisulfite solution be added to the mixture after the reaction?

(iii) What is the role of anhydrous sodium carbonate in the isolating procedure of cyclohexanone?

## 3.16 Synthesis of cinnamic acid

### 3.16.1 Purpose

(i) Understand the principle and method of cinnamic acid preparation.

(ii) Master the principle, application, and operation method of steam distillation.

### 3.16.2 Principle

Aromatic aldehydes and acetic anhydride can undergo a reaction similar to aldol condensation under base conditions to produce α, β-unsaturated aromatic aldehydes. This reaction is called the Perkin reaction. The catalyst is usually the potassium or sodium salt of the carboxylic acid of the corresponding acid anhydride. Potassium carbonate or tertiary amines also can be used as a catalyst for this reaction.

$$\text{PhCHO} + (CH_3CO)_2O \xrightarrow{CH_3COOK} \text{PhCH=CHCOOH} + CH_3COOH$$

### 3.16.3 Reagents

| | |
|---|---|
| Benzaldehyde | 10.5g |
| Acetic acetate | 15g |

Potassium acetate                6g
Sodium carbonate                20g
Concentrate hydrochloric acid

### 3.16.4  Apparatus

The schematic diagram of synthesis of cinnamic acid apparatus is shown in Fig 3-31.

### 3.16.5  Procedure

( i ) Place 10mL of benzaldehyde, 14mL of acetic anhydride, and 6g of finely powdered potassium acetate into a dry 250mL round-bottom flask.

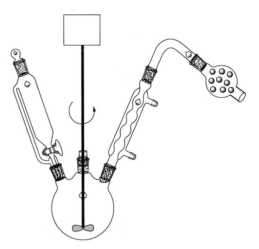

Fig 3-31  The schematic diagram of synthesis of cinnamic acid apparatus

( ii ) Set up the reflux device, connect a drying tube to the upper end of the condenser, and then heated to 160℃ for 60 minutes, and then increase it to 170-180℃ for about 3 hours.

( iii ) When the temperature decreases to 90-100℃, pour the reaction mixture into a 500mL round-bottomed flask that contains about 50mL of water. The resulting solution in the 500mL round-bottom flask is made alkaline by gradually adding a saturated sodium carbonate solution with vigorous shaking.

( iv ) The solution is subjected to steam distillation until all the "unreacted benzaldehyde" is removed. After the mixture cools in the distillation flask, activated carbon is added and heat to boiling, filtered by suction to remove unwanted by-products.

( v ) The filtrate is acidified by adding concentrated HCl gradually and shaking the flask until carbon dioxide evolution ceases completely.

( vi ) Cool the resulting solution with ice water, and a colorless solid precipitate appears; filter and wash it with a few amount of cold water, dry at 100℃.

### 3.16.6  Chart

The experimental process of synthesis of cinnamic acid is shown in Fig 3-32.

### 3.16.7  Notes

( i ) Acetic anhydride may be converted to acetic acid by hydrolysis when it is

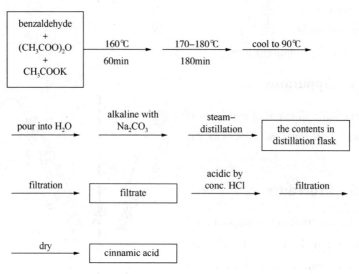

Fig 3-32  The experimental process of synthesis of cinnamic acid

exposed to moisture for a long time, so it should be redistilled before use as well.

(ⅱ) Aqueous sodium bicarbonate cannot be replaced by sodium hydroxide solution in the reaction.

(ⅲ) The Cinnamic acid can be recrystallized from other organic solvents.

### 3.16.8  Questions

(ⅰ) What kind of aldehyde can be used as starting compound for the Perkin condensation?

(ⅱ) Why cannot the aqueous sodium carbonate solution be replaced by a sodium hydroxide solution to neutralize the aqueous solution?

(ⅲ) What component is removed from the crude product by steam distillation? Can the crude product be purified by other methods instead of steam distillation?

## 3.17  Synthesis of acetylsalicylic acid

### 3.17.1  Purpose

(ⅰ) Learn the principle and method of preparing acetylsalicylic acid with acetic anhydride and salicylic acid under acid catalysis.

(ⅱ) Master the basic operations of reflux, recrystallization, and suction filtration.

## 3.17.2 Principle

$$\text{salicylic acid} + (CH_3CO)_2O \xrightarrow{H_2SO_4} \text{acetylsalicylic acid} + CH_3COOH$$

side reaction:

$$n \ \text{salicylic acid} \xrightarrow{H_2SO_4} \text{polymer}$$

## 3.17.3 Reagents

| | |
|---|---|
| Salicylic acid | 3.0g |
| Acetic anhydride | 6mL |
| Concentrate sulfuric acid | |
| Concentrate hydrochloride acid | |
| Saturated sodium bicarbonate solution | |

## 3.17.4 Apparatus

The schematic diagram of synthesis of acetylsalicylic acid apparatus is shown in Fig 3-33.

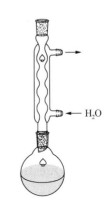

Fig 3-33  The schematic diagram of synthesis of acetylsalicylic acid apparatus

## 3.17.5 Procedure

( i ) Place 3.0g of salicylic acid into a 100mL Erlenmeyer flask, add 6mL of acetic anhydride and five drops of concentrated $H_2SO_4$; shake the mixture well and heat the flask to 80-90°C for 10 min.

( ii ) Remove the Erlenmeyer flask from the heat source and allow it to cool to room temperature. Add 40mL water and let the sample crystallize in an ice-water bath. Filter and wash the crystals with a small amount of cold water.

( iii ) Transfer the filter cake to a 150mL beaker, add 25mL of saturated sodium bicarbonate solution, and stir until no bubble of carbon dioxide is generated.

( iv ) Filter and wash the filter cake with 10mL of water, transfer the filtrate to a beaker, and slowly add concentrated hydrochloric acid to a pH of 1−2, and acetylsalicylic acid will precipitate out. Cool with ice water to complete the crystallization, filter and wash the precipitation with a small amount of cold water.

### 3.17.6 Chart

The experimental process of synthesis of acetylsalicylic acid is shown in Fig 3−34.

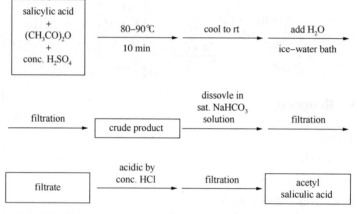

Fig 3−34  The experimental process of synthesis of acetylsalicylic acid

### 3.17.7 Notes

( i ) Before the experiment, the glassware needs to be dried. The acetic anhydride needs to be distilled before use, and collect the 139−140℃ fraction.

( ii ) If the reaction temperature is too low, the reaction will be insufficient; however, a high reaction temperature will increase the yield of by-products.

### 3.17.8 Questions

( i ) What is the role of concentrated sulfuric acid in the reaction of salicylic acid and acetic anhydride?

( ii ) What are the by-products that can be formed in this experiment?

## 3.18  Synthesis of 2-methyl-2-butanol

### 3.18.1 Purpose

( i ) Learn the preparation and application of Grignard reagent.

( ii ) Master the treatment methods of low-boiling flammable liquids.

## 3.18.2 Principle

In anhydrous diethyl ether, haloalkane reacts with magnesium to form alkyl magnesium halide, and the resulting alkyl magnesium halide reacts with ketone by addition-hydrolysis to obtain tertiary alcohols.

$$C_2H_5Br + Mg \xrightarrow{(C_2H_5)_2O} C_2H_5MgBr$$

$$\underset{H_3C\phantom{xx}CH_3}{\overset{O}{\|}} + C_2H_5MgBr \longrightarrow \underset{H_3C\phantom{xx}OMgBr}{\overset{H_3C\phantom{xx}C_2H_5}{\times}} \xrightarrow{H_3O^+} \underset{H_3C\phantom{xx}OH}{\overset{H_3C\phantom{xx}C_2H_5}{\times}}$$

The reaction should perform in the absence of water and oxygen because the Grignard reagent can be hydrolyzed with water. Thus, anhydrous ether is used as a solvent in this experiment. The preparation of Grignard reagent is an exothermic reaction, so it is necessary to control the rate of adding bromoethane, not too fast, and keep the solution boiling slightly. Acidic hydrolysis of the Grignard reagent with ketone admixture is also exothermic; thus it is better to perform the reaction under cooling conditions.

## 3.18.3 Reagents

| | |
|---|---|
| Ethyl bromide | 20mL (0.26mol) |
| Magnesium tape | 3.5g (0.144mol) |
| Acetone | 10mL (0.136mol) |

Diethyl ether

Iodine

Concentrated sulfuric acid

10% Sodium carbonate solution

Anhydrous potassium carbonate

## 3.18.4 Apparatus

The schematic diagram of synthesis of 2-methyl-2-butanol apparatus is shown in Fig 3-35.

## 3.18.5 Procedure

( i ) Place 3.5g of clean and dry magnesium

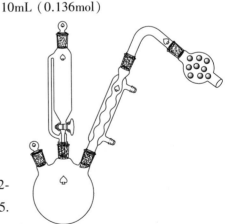

Fig 3-35 The schematic diagram of synthesis of 2-methyl-2-butanol apparatus

tape (cut into approximately 0.5cm segments) and 20mL of anhydrous ether in a 250mL three-necked flask; add 15mL of anhydrous ether and 20mL of bromoethane into the dropping funnel, shake well.

(ii) Start stirring, and add about 5-7mL of a mixture of bromoethane and ether from the dropping funnel to the three-necked flask.

(iii) When the solution boils slightly, and the color becomes gray and turbid, it indicates that the reaction has started (If there is no starting phenomenon in 10min, you can warm the flask with the palm of your hand or warm it in a warm water bath, or put in a small amount of iodine).

(iv) After confirming that the reaction has started, slowly add the mixture of bromoethane and ether, and adjust the dropping rate to keep the reaction flask boiling slowly (If the reaction is violent, the dropping should be suspended and cooled with cold water).

(V) After the dropping of bromoethane is finished, wait a few minutes, and then heat the flask slightly. After about 30min, magnesium is almost consumed, and the Grignard reagent is ready for use.

(vi) Cool the solution of ethylmagnesium bromide solution with a cold-water bath. Then a mixture of 10mL of anhydrous acetone and 10mL of anhydrous ether was slowly added with stirring. After the addition is completed, continue stirring for 5min.

(vii) Carefully add a pre-prepared mixture of 6mL concentrated sulfuric acid and 90mL water to the above solution. A white precipitate appears, and then the precipitate is dissolved again. Take care of the rate of adding, you should add slowly at the beginning.

(viii) Pour the mixture into a separatory funnel, and pour out the organic phase. Keep the water layer, and wash the ether layer with 15mL of 10% sodium carbonate solution. The ether layer is separated and retained. Then combine the alkaline layer and the initially retained aqueous layer, extract with 10mL diethyl ether for two times. The extracted ether layer is combined into the previously retained ether layer.

(ix) The combined ether layer was dried by adding anhydrous potassium carbonate, stoppered, and shaken until it became clear and transparent. The dried liquid is evaporated with a rotary evaporator to remove the solvent.

(X) Pour the remaining liquid into a 50mL round-bottom flask, add several zeolites, and collect 100-104℃ fractions by distillation.

### 3.18.6  Chart

The experimental process of synthesis of 2-methyl-2-butanol is shown in Fig 3-36.

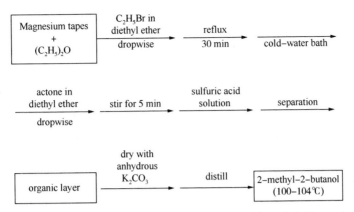

Fig 3-36  The experimental process of synthesis of 2-methyl-2-butanol

### 3.18.7  Notes

( i ) Commercially available magnesium strips should be treated in advance, and the oxide layer can be washed off with dilute hydrochloric acid.

( ii ) If the solution boils too violently, bromoethane will escape from the flask due to the low boiling point (38.4℃). Therefore, the dropping rate of bromoethane must be controlled strictly.

( iii ) Grignard reagent can react with oxygen, moisture, and carbon dioxide in the air, so the obtained solution of ethylmagnesium bromide should not be left for a long time, and the addition reaction should be done immediately.

( iv ) 2-Methyl-2-butanol can form a constant boiling mixture with water with a boiling point of 87.4℃. If the solution is not thoroughly dried and treated, some liquid will be distilled below 95℃, requiring re-drying and distillation.

### 3.18.8  Questions

( i ) When preparing Grignard reagents and performing nucleophilic addition reactions, why do all solvents and glassware should be dry absolutely?

( ii ) If the reaction does not proceed immediately, what kind of method can start the reaction? If the reaction does not begin, really, but you already add a large amount of bromoethane to the reaction, how to treat it?

( iii ) What kinds of desiccants do you know? Try to describe their functions and application range. Why can't the crude product obtained in this experiment be dried with calcium chloride?

( iv ) What are the advantages of using a rotary evaporator?

## 3.19 Synthesis of *tert*-butylhydroquinone

### 3.19.1 Purpose

( i ) Learn the principle and method of preparing *o-tert*-butylhydroquinone.

( ii ) Master experimental operations such as electric stirring, reflux, and recrystallization.

### 3.19.2 Principle

Hydroquinone is selected as a starting material for the synthesis of *tert*-butylhydroquinone. Usually, isobutylene or *tert*-butanol is used as an alkylation reagent. The common catalysts for this type of reaction include liquid catalysts and solid catalysts. Liquid catalysts include concentrated sulfuric acid, phosphoric acid, benzenesulfonic acid, and so on. The reaction is generally carried out in a mixture of water with organic solvents. Solid catalysts include strong acid ion exchange resins (such as Amberlyst-15), zeolites, and activated clay. The reaction needs to be carried out in solvents such as cycloalkanes, aromatic hydrocarbons, and aliphatic ketones.

In this experiment, hydroquinone reacts with *tert*-butanol in the presence of phosphoric acid in xylene to give *tert*-butyl hydroquinone.

The reaction proceeds in two steps: the first step is producing water-soluble intermediate-ethers, and the rate for this step is very fast; the second step is rearranging the intermediate to produce *ortho-tert*-butylhydroquinone. The second step is difficult, and it takes a long time to fully convert the intermediate to the desired product at a high temperature, which is the control step of the entire synthesis reaction.

## 3.19.3 Reagents

| | |
|---|---|
| tert-Butanol | 7.5mL (0.08mol) |
| Hydroquinone | 5.5g (0.05mol) |
| 85% Phosphoric acid | 5.0mL (0.075mol) |
| Xylene | 50mL (0.41mol) |

## 3.19.4 Apparatus

The schematic diagram of synthesis of tert-butylhydroquinone apparatus is shown in Fig 3-37.

## 3.19.5 Procedure

( i ) Install a two-neck connecting tube on a 150mL three-necked flask, then install a stirrer, thermometer, and reflux condenser. Add 5.5g of hydroquinone, 5.0mL of 85% phosphoric acid, and 20.0mL of xylene to the three-necked flask, and start stirring.

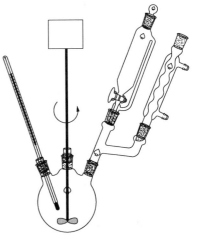

Fig 3-37 The schematic diagram of synthesis of tert-butylhydroquinone apparatus

( ii ) Slowly heat to 100-110℃, and slowly add 7.5mL tert-butanol (dissolve in 5mL xylene). During the dropping process, keep the temperature at 100-110℃, and start timing, about 30-60min, the dropping is finished.

( iii ) After the addition, continue to heat to 135-140℃, and reflux at constant temperature for 2.5h (time from when the tert-butanol is added).

( iv ) Slowly lower the temperature to about 120℃. When there is no reflux liquid, stop stirring and quickly transfer the reaction liquid into a beaker containing 50mL of hot water while it is hot. Wash the residual reaction liquid in the three-necked flask with a small amount of hot water, and combine it into the beaker.

( v ) Cool the beaker for about 30min to make it crystallize completely. Suction filtration yields a white crude product. The filtrate is separated and recovered xylene and phosphoric acid.

( vi ) Recrystallize with 25mL xylene and decolorize with activated carbon, suction filtration yields the desired product tert-butylhydroquinone

### 3.19.6 Chart

The experimental process of synthesis of *tert*-butylhydroquinone is shown in Fig 3-38.

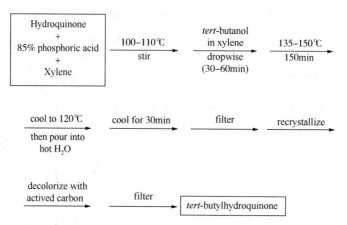

Fig 3-38  The experimental process of synthesis of *tert*-butylhydroquinone

### 3.19.7 Notes

( i ) When performing a heterogeneous reaction, or when one of the reactants is added by dropwise, stirring must be carried out to avoid local over-concentration and overheating leads to side reactions or decomposition of organic compounds.

( ii ) The purposes of xylene are: one is to control the concentration of *tert*-butanol to reduce the formation of by-product di-*tert*-butylhydroquinone; the other is to remove di-*tert*-butylhydroquinone in the product, which plays a primary purification role in the product.

### 3.19.8 Questions

( i ) What are the commonly used catalysts for Friedel-Crafts alkylation reaction?
( ii ) What are the benefits of using xylene as the solvent in this experiment?

## 3.20  Synthesis of nitrobenzene

### 3.20.1 Purpose

( i ) Learn the principle of the electrophilic substitution reaction on the benzene ring.

(ⅱ) Master experimental operations such as condensate reflux and water-bath heating.

## 3.20.2 Principle

Aromatic nitro compounds are generally prepared by direct nitration of aromatic compounds. The most commonly used nitrating reagent is a mixture of concentrated nitric acid and concentrated sulfuric acid, often called mixed acid.

$$\text{C}_6\text{H}_6 + \text{HNO}_3 \xrightarrow{\text{conc.H}_2\text{SO}_4} \text{C}_6\text{H}_5\text{NO}_2$$

## 3.20.3 Reagents

| | |
|---|---|
| Concentrated nitric acid | 3.6mL |
| Concentrated sulfuric acid | 5.0mL |
| Benzene | 4.5mL |
| Saturated sodium chloride solution | |
| Anhydrous calcium chloride | |

## 3.20.4 Apparatus

The schematic diagram of synthesis of nitrobenzene apparatus is shown in Fig 3-39.

## 3.20.5 Procedure

(ⅰ) Add 3.6mL of concentrated nitric acid into the Erlenmeyer flask and take another 5mL of concentrated sulfuric acid, add them to the Erlenmeyer flask several times, and shake them up while adding.

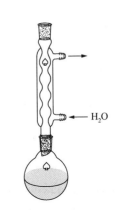

Fig 3-39  The schematic diagram of synthesis of nitrobenzene apparatus

(ⅱ) Add 4.5mL of benzene and the mixed acid prepared above into the flask, add several zeolites, shake well, mix well and start heating, control the water-bath temperature at about 60℃, and keep refluxing for 30min.

(ⅲ) The reaction mixture is poured into a separatory funnel to separate the product. The product is then placed in a conical flask and washed with an equal volume of water until it is not acidic, and finally washed with water until it is neutral. The obtained organic layer is placed in a dry conical flask and dried with anhydrous calcium chloride, and the volume of the product is finally measured. (Note: when washing with an equal volume of water for the first time, the organic layer is on the upper layer, and when washing with water for the second time, the organic layer is on the lower layer)

### 3.20.6 Chart

The experimental process of synthesis of nitrobenzene is shown in Fig 3-40.

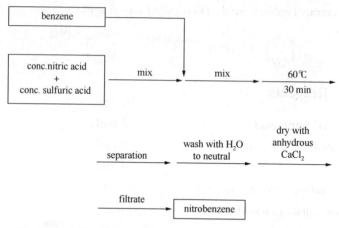

Fig 3-40  The experimental process of synthesis of nitrobenzene

### 3.20.7 Notes

( i ) The nitrification reaction is exothermic. The higher the temperature, the faster the speed of the nitration reaction, the more heat is released, and the more likely it is to cause temperature out of control and explosion.

( ii ) Most nitrified substances are flammable substances, and some are both toxic, such as benzene, toluene, etc. Improper use or storage can easily cause fires.

( iii ) Mixed acid has strong oxidizing and corrosive, and it can cause combustion when it comes in contact with organic matter, especially unsaturated organic compounds. The nitrification reaction is very corrosive and will cause strong corrosion of the equipment. During the preparation of the mixed acid, if the temperature is too high or a small amount of water is dropped, it will promote the decomposition of a large amount of nitric acid, causing sudden boiling or explosion.

( iv ) Most nitrification products have fire and explosion hazards, especially polynitro compounds and nitrate esters, which are very easy to explode or catch fire when exposed to heat, friction, impact, or contact with fire sources.

( v ) When preparing mixed acid, strictly control the temperature and ensure sufficient stirring and cooling conditions to prevent flushing or explosion caused by sharp temperature rise. Do not mix undiluted concentrated sulfuric acid with nitric acid. When diluting concentrated sulfuric acid, do not pour water into the acid.

(vi) When preparing a mixture of nitric acid and sulfuric acid, add sulfuric acid to the nitric acid in batches, and shake well while making it uniform.

(vii) During the nitrification reaction, the adding rate of mixed acid and the reaction temperature should be controlled strictly. The reaction system should have a good stirring and cooling device, and no water or power failure should occur in the middle of the system, and the stirring system should not fail. When the reaction temperature rises abnormally or the stirring fails, the feeding should be stopped and discharged to a safe place urgently.

(viii) The nitration reaction is exothermic. If the temperature exceeds 60℃, the by-product of dinitrobenzene is formed. If the temperature exceeds 80℃, benzenesulfonic acid is formed as a by-product, and some nitric acid and benzene volatilize away.

(ix) Nitro compounds are highly toxic to the human body. Inhalation of a large amount of steam or being absorbed by skin contact can cause poisoning! Therefore, you must be careful when handling nitrobenzene or other nitro compounds. If you accidentally touch your skin, you should immediately scrub with a small amount of ethanol and then wash with soap and warm water.

(X) When washing nitrobenzene, especially when washing with sodium carbonate solution, do not shake it too vigorously, otherwise the product will emulsify, and it will not be easy to layer. If this happens, add a few drops of alcohol and let it stand for a while to separate layers.

### 3.20.8 Questions

(i) What is the role of concentrated sulfuric acid in this experiment?
(ii) What will be happened if the temperature is too high during the reaction?
(iii) Why should the crude product be washed with water to neutralize?

## 3.21 Synthesis of adipic acid

### 3.21.1 Purpose

(i) Learn the principle and method of using cyclohexanol to prepare adipic acid.
(ii) Learn how to separate solid and liquid substances by suction filtration.

### 3.21.2 Principle

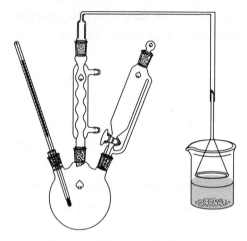

### 3.21.3 Reagents

| | |
|---|---|
| Cyclohexanol | 4.2mL |
| Concentrated nitric acid | 10mL |

### 3.21.4 Apparatus

The schematic diagram of synthesis of adipic acid apparatus is shown in Fig 3-41.

### 3.21.5 Procedure

( i ) Add 10mL of water and 10mL of concentrated nitric acid to the flask, mix well, and heat to 80℃ using a water bath. Add 4.2mL of cyclohexanol to the constant pressure dropping funnel. Control the dropping rate to keep the mixture's temperature in the flask between 85-90℃ (if necessary, add cold water to the water bath). After the addition of alcohol is finished, the reaction mixture should keep at 85-90℃ for another 15min to react fully.

Fig 3-41 The schematic diagram of synthesis of adipic acid apparatus

( ii ) Cool the reaction mixture in an ice-water bath, and filter the precipitated crystals through a Buchner funnel. Wash the filter cake with 3mL of water, and press the crystals as dry as possible and weigh them.

### 3.21.6 Chart

The experimental process of synthesis of adipic acid is shown in Fig 3-42.

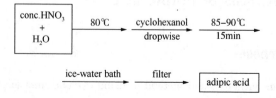

Fig 3-42 The experimental process of synthesis of adipic acid

## 3.21.7 Notes

(ⅰ) Cyclohexanol and nitric acid cannot be measured using the same measuring cylinder.

(ⅱ) The nitrogen dioxide gas produced in the experiment is toxic, so the device must be tight and airtight, and the exhaust gas must be absorbed.

(ⅲ) After the reaction is finished, the reaction solution should be poured out as soon as possible while the solution is hot. If cool to room temperature, adipic acid will crystallize out, and it is not easy to pour out, resulting in a loss of the desired product.

(ⅳ) Controlling the dropping rate of cyclohexanol is the key to the experiment of preparing adipic acid.

(ⅴ) The preparation of carboxylic acid adopts relatively strong oxidation conditions, which are generally exothermic reactions. The temperature for this reaction should be controlled strictly, otherwise it will affect the yield, but sometimes an explosion may occur.

## 3.21.8 Questions

(ⅰ) Why must the temperature of the oxidation reaction be strictly controlled?

(ⅱ) What should you do if the temperature in the flask exceeds 90℃?

# 3.22 Synthesis of dimethyl adipate

## 3.22.1 Purpose

(ⅰ) Learn the principle of esterification reaction and the preparation method of dimethyl adipate.

(ⅱ) Master experimental operations such as distillation and washing.

## 3.22.2 Principle

$$\text{C}_6\text{H}_{10}(\text{COOH})_2 + 2\text{CH}_3\text{OH} \xrightarrow[\Delta]{\text{H}_2\text{SO}_4(\text{cat.})} \text{C}_6\text{H}_{10}(\text{COOCH}_3)_2 + 2\text{H}_2\text{O}$$

## 3.22.3 Reagents

| | |
|---|---|
| Adipic acid | 5.0g |
| Methanol | 23mL |

Conc.$H_2SO_4$　　　　　　　　　　4mL

Toluene　　　　　　　　　　　　15mL

Diethyl ether

Anhydrous $MgSO_4$

Sat.$Na_2CO_3$ solution

Sat.NaCl solution

### 3.22.4 Apparatus

The schematic diagram of synthesis of dimethyl adipate apparatus is shown in Fig 3-43.

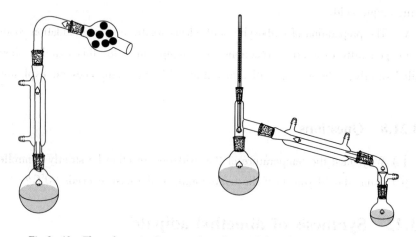

Fig 3-43　The schematic diagram of synthesis of dimethyl adipate apparatus

### 3.22.5 Procedure

( i ) Add 5.0g of adipic acid, 23mL of methanol (molar ratio of methanol to adipic acid = 20:1), and several zeolites into a round-bottomed flask, and then add 4mL of concentrated sulfuric acid in several times.

( ii ) Set up the reflux device, connect a drying tube to the upper end of the condenser, heat to reflux for 2h.

( iii ) After cooling, set up an atmospheric distillation device, distill most of the methanol under atmospheric pressure, then add 20g crushed ice, and extract with diethyl ether (15mL×3) 3 times, then combine the ether extracts.

( iv ) Wash the ether extracts with saturated sodium carbonate solution (20mL), saturated NaCl solution (10mL×3), and then dry with anhydrous magnesium sulfate (If magnesium sulfate does not stick to the wall of the flask, that means dry thoroughly).

(V) Filter to remove magnesium sulfate, distill the ether firstly, then distill methanol under reduced pressure to obtain a colorless liquid, weigh and calculate the yield.

### 3.22.6 Chart

The experimental process of synthesis of dimethyl adipate is shown in Fig 3-44.

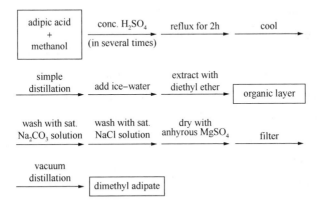

Fig 3-44  The experimental process of synthesis of dimethyl adipate

### 3.22.7 Notes

The preparation of dimethyl adipate (esterification reaction) is reversible, so the amount of methanol is excessive in this experiment, otherwise the adipic acid does not react fully, the utilization rate of adipic acid is low, and the output of by-product monomethyl adipate will be significantly increased.

### 3.22.8 Questions

(i) Can concentrated sulfuric acid be added at one time in this experiment?

(ii) In addition to concentrated sulfuric acid, what kind of catalysts can be used in this experiment?

## 3.23  Comprehensive experiment: green synthesis of 2,2'-furil

### 3.23.1 Purpose

(i) Cultivate students' ability to consult literature.

(ⅱ) Cultivate awareness of raw material purification.

(ⅲ) Stimulate students' innovative consciousness and spirit.

(ⅳ) Cultivate the ability to analyze product purity and expand students' knowledge.

### 3.23.2 Principle

From the perspective of green chemistry, this experiment uses furfural derived from biomass as a raw material to synthesize a high-value-added chemical 2,2'-furil in an efficient and environmentally friendly way. It enables students to understand the use of renewable energy and enhances students' good environmental protection and innovation awareness. By studying the purification of furfural, steam distillation of organic compounds, synthesis of furoin from furfural using vitamin $B_1$ as a catalyst, students can understand the concept of green catalysis. The third step is to synthesize octamolybdate quaternary ammonium salt, which is used as a catalyst in the oxidation of furoin under air conditions, which saves costs and avoids the generation of wastes. The experiment develops a green and efficient catalytic system, and introduces new concepts, new methods, and new technologies.

Furoin, the self-condensation product of furfural, cannot be produced on a large scale by traditional methods because the catalyst is cyanide, which cannot achieve the purpose of green chemistry. At present, use vitamin $B_1$($VB_1$ for short) as a catalyst, the materials are readily available, and the operation is relatively safe.

Today, furoin can be oxidized directly to produce 2,2'-furil with different oxidants, and the yields are relatively high. Typical catalysts include nitric acid, ferric chloride, manganese dioxide, barium manganate. These oxidants have a good oxidation effect on furoin, and the yield can reach about 90%. However, the above catalysts have related problems such as high costs and environmental pollution. In this experiment, a molybdate catalyst was synthesized, which can effectively catalyze the conversion of furoin to 2,2'-furil under air conditions, avoiding the use of strong oxidants and metal oxidants.

ⅰ) preparation of the catalyst

$$4\,\underset{\underset{CH_2CH_2CH_2CH_3}{|}}{\overset{+}{Py}}\,Br^- + 8Na_2MoO_4 + 12HCl \longrightarrow \left[\underset{\underset{CH_2CH_2CH_2CH_3}{|}}{\overset{+}{Py}}\right]_4 Mo_8O_{26}$$

$$[PyC_4H_9]_4Mo_8O_{26}$$

ii) oxidation of furoin

[structural reaction: furoin $\xrightarrow{[PyC_4H_9]_4Mo_8O_{26}}$ 2,2'-furil]

### 3.23.3 Reagents

|  |  |
|---|---|
| Furfural | 15 mL |
| 1-Bromobutane | 4 mL |
| Pyridine | 2 mL |
| Na$_2$MoO$_4$ | 1.2 g |
| VB$_1$ | 1.5 g |
| Acetic acid | 10 mL |
| 10% NaOH solution | |
| Ethanol | |

### 3.23.4 Apparatus

The experimental apparatus involved in the green synthesis of 2,2'-furil is shown in Fig 3-45.

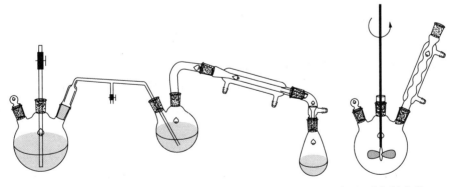

Fig 3-45  The experimental apparatus involved in the green synthesis of 2,2'-furil

### 3.23.5 Procedure

(a) Synthesis of furoin

ⅰ) Purification of furfural

Add 15mL of furfural and 20mL of water into a 100mL round bottom flask containing several zeolites, install a distillation device, and heat it with an electric heater. When the temperature reaches 100℃, steam begins to appear. Control the temperature, receive the colorless furfural aqueous solution, separate and dry it for later use.

ⅱ) Synthesis of furoin

In a three-necked flask equipped with an electric stirrer, reflux condenser and thermometer, add 1.5g of $VB_1$, 3mL of water, and 8mL of 95% ethanol. The solution is colorless. Then add 10% NaOH solution, adjust the pH to 9-10 (about 2mL). After the addition, the color of the solution changes from colorless to reddish-brown. Then add 4.8mL freshly distilled furfural, and keep the pH to 9-10 and the temperature at 65-70℃. After adding, transfer the solution to 200mL of ice water, crystallize the product, filter and wash with ice water to give a light yellow solid. This solid is further recrystallized with 95% ethanol and dried for later use.

(b) Preparation of quaternary ammonium salt catalyst

ⅰ) Preparation of butyl pyridinium bromide quaternary ammonium salt

Place 2mL pyridine and 4mL 1-bromobutane in a dry 50mL round bottom flask, add several zeolites, heat to reflux for one hour. Cool the reaction solution, and add 10mL distilled water and several zeolites into the flask, distill and collect the distillate until there are no oil droplets. The residue in the round-bottom flask (n-butyl pyridine bromide aqueous solution) is for later use.

ⅱ) Preparation of tetrabutylpyridinium octamolybdate catalyst

Place 1.2g of sodium molybdate dihydrate in a beaker, add 10mL of distilled water, and slowly add 2mL of dilute hydrochloric acid (concentration in 1:3 by volume, that is, take the mixture of concentrated hydrochloric acid and two volumes of distilled water), constantly stir to dissolve completely and obtain a colorless and transparent solution.

Take half of the n-butyl pyridine bromide aqueous solution prepared above and add it to the sodium molybdate solution under stirring. A large amount of precipitation is formed immediately. Stir to disperse the resulting material, filter under reduced pressure, wash it with a small amount of distilled water, and dry for later use.

(c) Synthesis of 2,2'-furil

First, add 0.5g of furoin (the mass ratio between the two is 5∶1) and 0.1g of the prepared catalyst into a three-necked flask equipped with an electric stirring and condensing reflux device. Then add 10mL of glacial acetic acid, and reflux at 110℃ for two hours. The solution gradually changed from blue to green, and finally yellow crystals appeared on the vessel's walls. After the reaction, cool the reaction mixture to room temperature, add 10mL of ice water, and then mix evenly to settle for precipitation. Filter and dry under vacuum at 70℃ for 24h to give the desired product.

### 3.23.6 Chart

(a) Synthesis of furoin

The experimental process of synthesis of furoin is shown in Fig 3-46.

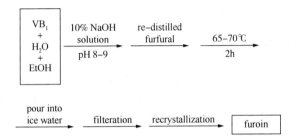

Fig 3-46  The experimental process of synthesis of furoin

(b) Preparation of $[PyC_4H_9]_4Mo_8O_{26}$

The experimental process of preparation of $[PyC_4H_9]_4Mo_8O_{26}$ is shown in Fig 3-47.

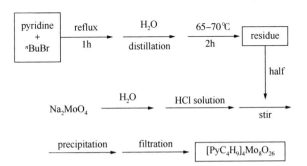

Fig 3-47  The experimental process of preparation of $[PyC_4H_9]_4Mo_8O_{26}$

(c) Oxidation of furoin to 2,2'-furil

The experimental process of oxidation of furoin to 2,2'-furil is shown in Fig 3-48.

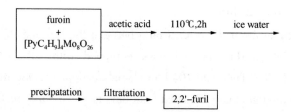

Fig 3-48 The experimental process of oxidation of furoin to 2,2'-furi

### 3.23.7 Notes

( i ) Furfural needs to be distilled again before use.

( ii ) Pyridine has a foul smell, so weighing needs to be carried out in a fume hood. If you feel uncomfortable during the experiment, you need to rest for a few minutes in a well-ventilated place.

( iii ) During the experiment, please pay attention to prevent burns.

### 3.23.8 Questions

( i ) Why not use simple distillation for re-distilled furfural?

( ii ) For the preparation of furoin, why the pH of the solution needs to be controlled between 9-10?

( iii ) What is the reaction mechanism using $VB_1$ as a catalyst?

( iv ) Is there any pathway to synthesize 2,2'-furil?

# Chapter 4

# Appendix

## 4.1 List of the element with their symbols and atomic masses

Table 4-1  The element with their symbols and atomic masses

| Element | Symbol | Atomic Number | Atomic Mass |
|---|---|---|---|
| Hydrogen | H | 1 | 1.008 |
| Helium | He | 2 | 4.003 |
| Lithium | Li | 3 | 6.941 |
| Beryllium | Be | 4 | 9.012 |
| Boron | B | 5 | 10.81 |
| Carbon | C | 6 | 12.01 |
| Nitrogen | N | 7 | 14.01 |
| Oxygen | O | 8 | 16.00 |
| Fluorine | F | 9 | 19.00 |
| Neon | Ne | 10 | 20.18 |
| Sodium | Na | 11 | 22.99 |
| Magnesium | Mg | 12 | 24.31 |
| Aluminum | Al | 13 | 26.98 |
| Silicon | Si | 14 | 28.09 |
| Phosphorus | P | 15 | 30.97 |
| Sulfur | S | 16 | 32.07 |
| Chlorine | Cl | 17 | 35.45 |
| Argon | Ar | 18 | 39.95 |
| Potassium | K | 19 | 39.10 |
| Calcium | Ca | 20 | 40.08 |

Continued

| Element | Symbol | Atomic Number | Atomic Mass |
|---|---|---|---|
| Scandium | Sc | 21 | 44.96 |
| Titanium | Ti | 22 | 47.88 |
| Vanadium | V | 23 | 50.94 |
| Chromium | Cr | 24 | 52.00 |
| Manganese | Mn | 25 | 54.94 |
| Iron | Fe | 26 | 55.85 |
| Cobalt | Co | 27 | 58.93 |
| Nickel | Ni | 28 | 58.69 |
| Copper | Cu | 29 | 63.55 |
| Zinc | Zn | 30 | 65.39 |
| Gallium | Ga | 31 | 69.72 |
| Germanium | Ge | 32 | 72.59 |
| Arsenic | As | 33 | 74.92 |
| Selenium | Se | 34 | 78.96 |
| Bromine | Br | 35 | 79.90 |
| Krypton | Kr | 36 | 83.80 |
| Rubidium | Rb | 37 | 85.47 |
| Strontium | Sr | 38 | 87.62 |
| Yttrium | Y | 39 | 88.91 |
| Zirconium | Zr | 40 | 91.22 |
| Niobium | Nb | 41 | 92.91 |
| Molybdenum | Mo | 42 | 95.96 |
| Technetium | Tc | 43 | (99) |
| Ruthenium | Ru | 44 | 101.1 |
| Rhodium | Rh | 45 | 102.9 |
| Palladium | Pd | 46 | 106.4 |
| Silver | Ag | 47 | 107.9 |
| Cadmium | Cd | 48 | 112.4 |
| Indium | In | 49 | 114.8 |
| Tin | Sn | 50 | 118.7 |
| Antimony | Sb | 51 | 121.8 |

Continued

| Element | Symbol | Atomic Number | Atomic Mass |
|---|---|---|---|
| Tellurium | Te | 52 | 127.6 |
| Iodine | I | 53 | 126.9 |
| Xenon | Xe | 54 | 131.3 |
| Cesium | Cs | 55 | 132.9 |
| Barium | Ba | 56 | 137.3 |
| Lanthanum | La | 57 | 138.9 |
| Cerium | Ce | 58 | 140.1 |
| Praseodymium | Pr | 59 | 140.9 |
| Neodymium | Nd | 60 | 144.2 |
| Promethium | Pm | 61 | (145) |
| Samarium | Sm | 62 | 150.4 |
| Europium | Eu | 63 | 152.0 |
| Gadolinium | Gd | 64 | 157.3 |
| Terbium | Tb | 65 | 158.9 |
| Dysprosium | Dy | 66 | 162.5 |
| Holmium | Ho | 67 | 164.9 |
| Erbium | Er | 68 | 167.3 |
| Thulium | Tm | 69 | 168.9 |
| Ytterbium | Yb | 70 | 173.1 |
| Lutetium | Lu | 71 | 175.0 |
| Hafnium | Hf | 72 | 178.5 |
| Tantalum | Ta | 73 | 180.9 |
| Tungsten | W | 74 | 183.8 |
| Rhenium | Re | 75 | 186.2 |
| Osmium | Os | 76 | 190.2 |
| Iridium | Ir | 77 | 192.2 |
| Platinum | Pt | 78 | 195.1 |
| Gold | Au | 79 | 197.0 |
| Mercury | Hg | 80 | 200.6 |
| Thallium | Ti | 81 | 204.38 |
| Lead | Pb | 82 | 207.2 |

Continued

| Element | Symbol | Atomic Number | Atomic Mass |
|---|---|---|---|
| Bismuth | Bi | 83 | 209.0 |
| Polonium | Po | 84 | (210) |
| Astatine | At | 85 | (210) |
| Radon | Rn | 86 | (222) |
| Francium | Fr | 87 | (223) |
| Radium | Ra | 88 | (226) |
| Actinium | Ac | 89 | (227) |
| Thorium | Th | 90 | 232.0 |
| Protactinium | Pa | 91 | 231.0 |
| Uranium | U | 92 | 238.0 |
| Neptunium | Np | 93 | (237) |
| Plutonium | Pu | 94 | (239) |
| Americium | Am | 95 | (243) |
| Curium | Cm | 96 | (247) |
| Berkelium | Bk | 97 | (247) |
| Califonium | Cf | 98 | (252) |
| Einsteinium | Es | 99 | (252) |
| Fermium | Fm | 100 | (257) |
| Mendelevium | Md | 101 | (258) |
| Nobelium | No | 102 | (259) |
| Lawrencium | Lr | 103 | (262) |
| Rutherfordium | Rf | 104 | (267) |
| Dubnium | Db | 105 | (268) |
| Seaborgium | Sg | 106 | (271) |
| Bohrium | Bh | 107 | (272) |
| Hassium | Hs | 108 | (277) |
| Meitnerium | Mt | 109 | (276) |
| Darmstadtium | Ds | 110 | (281) |
| Roentgenium | Rg | 111 | (280) |
| Copernicium | Cn | 112 | (285) |
| Nihonium | Nh | 113 | (284) |

Continued

| Element | Symbol | Atomic Number | Atomic Mass |
|---|---|---|---|
| Flerovium | Fl | 114 | (289) |
| Moscovium | Mc | 115 | (288) |
| Livermorive | Lv | 116 | (293) |
| Tennessine | Ts | 117 |  |
| Oganesson | Og | 118 | (294) |

## 4.2 Structures and names for common solvents used in organic chemistry

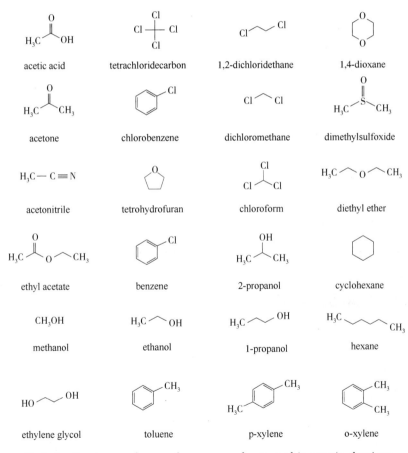

Fig 4-1  Structures and names for common solvents used in organic chemistry

## 4.3 Properties for common solvents used in organic chemistry

Table 4-2 The properties for common solvents used in organic chemistry

| Solvent | Formula | MW | Boiling point/℃ | Melting point/℃ | Density/ (g/mL) | Solubility in water/ (g/100g) |
|---|---|---|---|---|---|---|
| Acetic acid | $C_2H_4O_2$ | 60.052 | 118 | 16.6 | 1.0446 | Miscible |
| Acetone | $C_3H_6O$ | 58.079 | 56.05 | -94.7 | 0.7845 | Miscible |
| Acetonitrile | $C_2H_3N$ | 41.052 | 81.65 | -43.8 | 0.7857 | Miscible |
| Benzene | $C_6H_6$ | 78.11 | 80.1 | 5.5 | 0.8765 | 0.18 |
| 1-Butanol | $C_4H_{10}O$ | 74.12 | 117.7 | -88.6 | 0.8095 | 6.3 |
| Chlorobenzene | $C_6H_5Cl$ | 112.56 | 131.7 | -45.3 | 1.1058 | 0.05 |
| Chloroform | $CHCl_3$ | 119.38 | 61.2 | -63.4 | 1.4788 | 0.795 |
| Cyclohexane | $C_6H_{12}$ | 84.16 | 80.7 | 6.6 | 0.7739 | 0.0055 |
| 1,2-Dichloroethane | $C_2H_4Cl_2$ | 98.96 | 83.5 | -35.7 | 1.245 | 0.861 |
| Diethyl ether | $C_4H_{10}O$ | 74.12 | 34.5 | -116.2 | 0.713 | 7.5 |
| Diglyme | $C_6H_{14}O_3$ | 134.17 | 162 | -68 | 0.943 | Miscible |
| 1-Propanol | $C_3H_8O$ | 60.10 | 97 | -126 | 0.803 | Miscible |
| Pyridine | $C_5H_5N$ | 79.10 | 115.2 | -41.6 | 0.982 | Miscible |
| Tetrahydrofuran | $C_4H_8O$ | 72.106 | 65 | -108.4 | 0.8833 | Soluble |
| Toluene | $C_7H_8$ | 92.14 | 110.6 | -93 | 0.867 | 0.05 |
| Triethyl amine | $C_6H_{15}N$ | 101.19 | 88.9 | -114.7 | 0.728 | 0.02 |
| Water | $H_2O$ | 18.02 | 100.00 | 0.00 | 0.998 | |

## 4.4 Cooling bath

Table 4-3 Cooling baths

| Temperature/℃ | Composition | Temperature/℃ | Composition |
|---|---|---|---|
| 13 | P-Xylene/$CO_2$(s) | -77 | Acetone/$CO_2$(s) |
| 6 | Cyclohexane/$CO_2$(s) | -83.6 | Ethyl Acetate/Lip $N_2$ |
| 2 | Formamide/$CO_2$(s) | -94 | Hexane/Lip $N_2$ |
| -5--20 | Ice/Salt | -95.1 | Toluene/Lip $N_2$ |
| -41 | Acetonitrile/$CO_2$(s) | -100 | Ethyl Ether/$CO_2$(s) |

Continued

| Temperature/℃ | Composition | Temperature/℃ | Composition |
|---|---|---|---|
| −116 | Ethyl Ether/Lip $N_2$ | −83 | Propyl Amine/$CO_2$(s) |
| −131 | n-Pentane/Lip $N_2$ | −89 | n-Butanol/Lip $N_2$ |
| −196 | Lip $N_2$ | −94.6 | Acetone/Lip $N_2$ |
| 12 | Dioxane/$CO_2$(s) | −98 | Methanol/Lip $N_2$ |
| 5 | Benzene/$CO_2$(s) | −104 | Cyclohexane/Lip $N_2$ |
| 0 | Crushed Ice | −116 | Ethanol/Lip $N_2$ |
| −10.5 | Ethylene Glycol/$CO_2$(s) | −160 | Isopentane/Lip $N_2$ |
| −42 | Pyridine/$CO_2$(s) | | |

## 4.5 Common azeotropes

### Table 4-4 Common azeotropes

| Azeotrope | Boiling point/℃ | Compsotion(w/w) | Azeotropic point/℃ |
|---|---|---|---|
| Water-acetonitrile | −81.5 | 14.2−85.7 | 76 |
| Water-benzene | −80.4 | 8.8−91.2 | 69.2 |
| Water-dichloroethane | −83.7 | 19.5−80.5 | 72.0 |
| Water-dioxane | −101.3 | 18−82 | 87.7 |
| Water-ethanol | 100−78.5 | 5−95 | 78.15 |
| Water-ethyl acetate | −78 | 9.0−91 | 70 |
| Water-formic acid | −101 | 26−74 | 107 |
| Water-isopropanol | 82.4 | 12.1−87.9 | 80.4 |
| Water-n-propand | −97.2 | 28.8−71.2 | 87.7 |
| Water-propanoic acid | −141.4 | 82.2−17.8 | 99.1 |
| Water-pyridyl | −115.5 | 42−58 | 94.0 |
| Water-1-pentanol | −138.3 | 44.7−55.3 | 95.4 |
| Water-tert-butanol | −82.5 | 11.8−88.2 | 79.9 |
| Water-toluene | −110.5 | 20−80 | 85.0 |
| Ethanol-benzene | −80.6 | 32−68 | 68.2 |
| Ethanol-chloroform | −61.2 | 7−93 | 59.4 |
| Ethanol-ethyl acetate | 78.3−78.0 | 30−70 | 72.0 |

## 4.6 Pressure-temperature nomograph

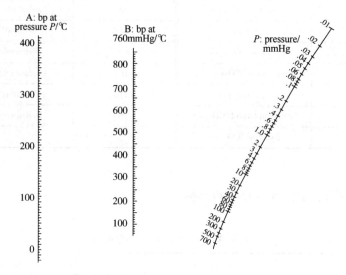

Fig 4-2  Pressure-temperature nomograph

## 4.7 Periodic table

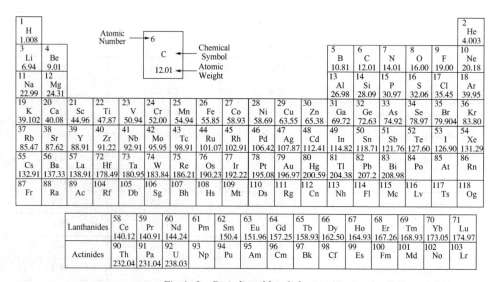

Fig 4-3  Periodic table of elements